Between Ape and Artilect

Conversations with Pioneers of Artificial General Intelligence and Other Transformative Technologies

Interviews Conducted and Edited by Ben Goertzel

This work is offered under the following license terms:

Creative Commons: Attribution-NonCommercial-NoDerivs 3.0 Unported (CC-BY-NC-ND-3.0)

See http://creativecommons.org/licenses/by-nc-nd/3.0/ for details

Copyright © 2013 Ben Goertzel

All rights reserved.

ISBN: 1496138171

ISBN-13: 9781496138170

"Man is a rope stretched between the animal and the Superman – a rope over an abyss."

-- Friedrich Nietzsche, *Thus Spake Zarathustra*

Table of Contents

Introduction ... 7
Itamar Arel: AGI via Deep Learning ... 11
Pei Wang: What Do You Mean by "AI"? ... 23
Joscha Bach: Understanding the Mind ... 39
Hugo DeGaris: Will There be Cyborgs? ... 51
DeGaris Interviews Goertzel: Seeking the Sputnik of AGI ... 61
Linas Vepstas: AGI, Open Source and Our Economic Future ... 89
Joel Pitt: The Benefits of Open Source for AGI ... 101
Randal Koene: Substrate-Independent Minds ... 107
João Pedro de Magalhães: Ending Aging ... 129
Aubrey De Grey: Aging and AGI ... 149
David Brin: Sousveillance ... 157
J. Storrs Hall: Intelligent Nano Factories and Fogs ... 173
Mohamad Tarifi: AGI and the Emerging Peer-to-Peer Economy ... 187
Michael Anissimov: The Risks of Artificial Superintelligence ... 203
Muehlhauser & Goertzel: Rationality, Risk, and the Future of AGI ... 227
Paul Werbos: Will Humanity Survive? ... 269
Wendell Wallach: Machine Morality ... 283
Francis Heylighen: The Emerging Global Brain ... 295
Steve Omohundro: The Wisdom of the Global Brain and the Future of AGI 325
Alexandra Elbakyan: Beyond the Borg ... 337
Giulio Prisco: Technological Transcendence ... 355
Zhou Changle: Zen and the Art of Intelligent Robotics ... 373
Hugo DeGaris: Is God an Alien Mathematician? ... 389
Lincoln Cannon: The Most Transhumanist Religion? ... 407
Natasha Vita-More: Upgrading Humanity ... 439
Jeffery Martin & Mikey Siegel: Engineering Enlightenment ... 457

Introduction

The amazing scientific and technological progress that characterizes our times, is at this stage still being directed primarily by flesh and blood human beings. Behind every radical scientific or technological advance or attempt, is some human being – or team -- with their own particular idiosyncratic perspective on their work and on the world.

Due to the various roles I've taken in the science, technology and futurist communities in recent years, I have come into contact with a remarkable variety of intriguing, transformational scientists, engineers and visionaries – people bringing their own individual, often wonderfully quirky perspectives to their work on revolutionary technologies and related concepts. I have dialogued with these pathbreaking people in various ways and contexts: face to face and on line, in groups and individually, in highly focused settings or over lunch or beer or while hiking in the mountains, etc.

In late 2010 I decided it would be interesting to formalize some of my ongoing dialogues with various interesting researchers, in the form of interviews/dialogues to be published in H+ Magazine, an online transhumanist magazine for which I have off-and-on written and/or edited since 2008. So I conducted a variety of such dialogues, mostly in late 2010 and early 2011 but continuing into 2012. What I did was to email each of my subjects a list of questions, get their responses via email, then reply to their responses, and so forth -- and then, finally, edit the results into a coherent linear dialogue. The resulting dialogues were well received when posted on the H+ Magazine website; and I eventually decided they have enough lasting value that it makes sense to collect them all into a book.

The most central focus of the dialogues in the book is Artificial General Intelligence (AGI) – the quest to make software or hardware with the same sort of thinking power as humans, and ultimately a lot more. Between half and two thirds of the chapters are focused in some way on AGI. But there's also plenty of variation from this focus: a couple chapters on life extension research, a chapter on sousveillance, one on nanotech, some chapters on the Global Brain and on transhumanist spirituality. AGI does rear its head now and then even in the chapters that don't focus on it, though. One of the exciting and sometimes mind-boggling characteristics of the current revolution in science and technology, is the way all the different branches of inquiry seem to be melding together into one bold, complex quest to understand and reshape the universe.

The online H+ Magazine versions of the dialogues, in some cases, have nice pictures that didn't make it into this book version. You may look the articles up online if you're curious! However, the book format has a different flavor: it puts all the dialogues together, allowing synergies between the different perspectives to be readily perceived; and it nudges the reader more toward thoughtful cogitation on the themes discussed, whereas Web pages tend more to encourage quick-and-dirty skimming.

Some of the material in some of the dialogues is already a bit dated – in some areas of science, a couple years is a long time! But I haven't updated or otherwise changed any of the material in the dialogues from the original H+ Magazine versions, except for formatting adjustments. On the other hand, I have more significantly tweaked the introductory prose prior to the start of some of the dialogues. The information about interviewees' careers and affiliations has been updated so as to be correct as of January 2013; and introductory information that was appropriate for the magazine but not for a book has been removed.

Only a few of the pieces given here have the flavor of classical "interviews" – in most cases they are more like "dialogues", meaning that I inject a lot of my own ideas and opinions. Except

Introduction

in the case of Hugo de Garis's interview of me (the "AGI Sputnik" dialogue given here), the focus is on the ideas of the other person I'm interviewing. But the feel in most cases is more of a wide-ranging conversation between friends/colleagues, rather than a reporter interviewing a subject. This is both because I like to talk too much, and because unlike most reporters, I myself happen to have a deep scientific expertise in the research areas being pursued by the subject.

In spite of the conversational nature of the material, this is not exactly "light reading" – the lay reader with no background in the various areas of science and technology being discussed, is bound to get lost here and there. However, it's certainly not *technical* material either. The discussion is kept at the level one might hear in a panel discussion at a non-technical technology-oriented conference like a Singularity Summit, Humanity+ conference or TEDx. If you like the idea of listening in on conversations between scientists and engineers talking about their work and their ideas, then this book is definitely for you!

The ideas discussed in these pages are not only a lot of fun to think about and work on, they are also among the most important things happening on the planet at this point in time. Furthermore, they are ideas that nobody understands all that thoroughly – in spite of their importance, and the relative speed at which they are transitioning from idea into reality. I think it's important that we foster a diversity of voices on these topics, and that folks with different opinions make every effort to understand each other and absorb each other's ideas. That is the spirit in which these dialogues were conducted, and in which I'm now presenting them to you in book form.

The title "Between Ape and Artilect" was chosen to reflect the transitional nature of our times. We are creatures of flesh and blood, on the verge of engineering creatures made from a much wider variety of substances, and of transmogrifying ourselves into all sorts of novel forms. We carry minds and bodies that are largely the same as those of apes; yet we are able to conceive, and soon will be able to create, artificial general intelligences with physical and mental capabilities that are, from the perspective of our current comprehension, essentially unlimited.

The word "artilect" was coined by Hugo DeGaris, who is involved in three of the dialogues contained here, as a short form of "artificial intellect."   I like the sound of it, but actually I find the etymology slightly problematic. An artifice is a tool, but will the cyborgs and superhuman AGI systems we create in the next phase of our civilization, really be anybody's tool?

Nietzsche insightfully conceived humanity as a rope between animal and superhuman – but he didn't have a clear picture of exactly how humanity was going to create the superhuman. Now, while plenty of technical details remain to be worked out, we can see how a lot more clearly.

The first dialogue given here, with AGI researcher Itamar Arel, focuses pretty squarely on the particulars of how to build human-like minds using current technologies. Itamar is an engineer and a very down-to-earth kind of visionary. The last few dialogues in the book are at the other end of the spectrum: looking at the potential of AGIs, neurofeedback, brain-computer interfacing and other technologies to mediate spiritual as well as technological and intellectual advancement.   The chapters between cover a variety of practical and theoretical topics.   But however far "out there" the contents get, everyone whom I've dialoguing with is a serious, deep thinker; and nearly all are experienced professional scientists. In these times, the border between science fictional and pragmatic thinking grows fuzzier and fuzzier – a fact that I find to be a heck of a lot of fun!

Itamar Arel: AGI via Deep Learning

Interview first published February 2011

Itamar Arel is one of those rare AI academics who has the guts to express open AGI optimism. He believes human-level AGI is achievable via deep learning and reinforcement learning, and that a properly funded AGI initiative could get there pretty quickly – potentially well ahead of Ray Kurzweil's proposed 2029 date.

Itamar's background spans business and academia. He runs the Machine Intelligence Lab at the University of Tennessee and also co-founded the Silicon Valley AI startup Binatix Inc[1]. Though he started his career with a focus on electrical engineering and chip design, Itamar has been pursuing AGI for many years now, and has created a well-refined body of theory as well as two bodies of proto-AGI software code. In recent years he's begun to reach out to the futurist community as well as the academic world, speaking at the 2009 Singularity Summit[2] in New York and the 2010 H+ Summit @ Harvard[3].

I discovered Itamar's work in 2008 with great enthusiasm, because I had always been interested in Jeff Hawkins' Hierarchical Temporal Memory approach to machine perception[4], but I'd been frustrated with some of the details of Hawkins' work. It appeared to me that Itamar's ideas about perception were conceptually somewhat similar to Hawkins's, but that Itamar had worked out many of the details in a more compelling way. Both Itamar's and Hawkins' approaches to perception involve hierarchical networks of processing elements,

---

[1] http://binatix.com
[2] http://vimeo.com/7318781
[3] http://hplussummit.com/arel.html
[4] http://www.numenta.com/

which self-organize to model the state and dynamics of the world. But it seemed to me that Itamar's particular equations were more likely to cause the flow of information up and down the hierarchy to cause the overall network to organize into an accurate world-model. Further, Itamar had a very clear vision of how his hierarchical perception network would fit into a larger framework involving separate but interlinked hierarchies dealing with actions and goals. (Another exciting project that has recently emerged in the same general space is that of Hawkins' former collaborator Dileep George[5], who has now broken off to start his own company, Vicarious Systems[6]. But I can't comment on the particulars of George's new work, as little has been made public yet.)

I had a chance to work closely with Itamar when we co-organized a workshop at University of Tennessee in October 2009, at which 12 AGI researchers gathered to discuss the theme of "A Roadmap Toward Human-Level AGI" (a paper summarizing the findings of that workshop has been submitted to a journal and should appear "any day now"). I'm also currently using his Machine Intelligence Lab's DeSTIN[7] software together with my team's OpenCog software, to supply OpenCog with a robot perception module. So as you can imagine, we've had quite a few informal conversations on the ins and outs of AGI over the last couple years. This interview covers some of the same ground but in a more structured way.

Ben
Before plunging into the intricacies of AGI research, there are a couple obvious questions about the AGI field as a whole I'd like to get your take on. First of all the ever-present funding issue.

One of the big complaints one always hears among AGI researchers is that there's not enough funding in the field. And

---

[5] http://www.dileepgeorge.com/
[6] http://www.vicariousinc.com
[7] http://web.eecs.utk.edu/~itamar/Papers/BICA2009.pdf

for sure, if you look at it comparatively, the amount of effort currently going into the creation of powerful AGI right now is really rather low, relative to the magnitude of benefit that would be obtained from even modest success, and given the rapid recent progress in supporting areas like computer hardware, computer science algorithms and cognitive science. To what do you attribute this fact?

Itamar

I feel that the answer to this question has to do with understanding the history of AI. When the term AI was first coined, over 50 years ago, prominent scientists believed that within a couple of decades robots will be ubiquitous in our lives, helping us with every day chores and exhibiting human-level intelligence. Obviously, that didn't happen yet. However, funding agencies, to some degree, have lost hope in the holy grail of AGI given the modest progress made toward this goal.

Ben

So the current generation of researchers is paying the price for the limited progress made by prior generations—even though now we have so much more understanding and so much better hardware.

Itamar

That's correct. Historical under-delivery on the promise of achieving AGI is probably the most likely reason for the marginal funding being directed at AGI research. Hopefully, that will change in the near future.

Ben

Some people would say that directing funding to AGI isn't so necessary, since "narrow AI" research (focusing on AI programs that solve very specific tasks and don't do anything else, lacking the broad power to generalize) is reasonably well funded. And of course the funding directed to narrow AI is understandable, since some narrow AI applications are delivering current value to many people (e.g. Internet search, financial and military applications, etc.). So, these folks would argue that AGI will eventually be achieved via incremental improvement of narrow-AI technology – i.e. that narrow AI can gradually become better and better by becoming broader and broader, until eventually it's AGI. So then there's no need for explicit AGI funding. What are your views on this?

Itamar

I think my view on this is fairly similar to yours. I believe the challenges involved in achieving true AGI go well beyond those imposed by any narrow AI domain. While separate components of an AGI system can be degraded and applied to narrow AI problem domains, such as a scalable perception engine (the focus of much of my current work), it is only in the integration and the scaling of the system as a whole that AGI can be demonstrated. From a pragmatic standpoint, developing pieces of AGI separately and applying them to problems that are more easily funded, is a valid approach.

Ben

But even if you do use narrow AI funding to develop pieces of your AGI system, there's still going to be a chunk of work that's explicitly focused on AGI, and doesn't fit into any of the narrow AI applications, right? This seems extremely obvious to me, and you seem to agree – but it's surprising that not everybody sees this.

Itamar

I think that due to the misuse of the term AI, the public is somewhat confused about what AI really means. The majority of the people have been lead to believe that rule-based expert systems, such as those driving modern games, are actually "AI". In general, I suspect there is little trust in the premise of AGI, to the point where people probably associate such technology with movies and fiction.

Ben

One AGI-related area that's even better-funded than narrow AI is neuroscience. So, one approach that has been proposed for surmounting the issues facing AGI research is to piggyback on neuroscience, and draw on our (current or future) knowledge of the brain. I'm curious to probe your views on this a bit

Itamar

While reverse – engineering the brain is, almost by definition, a worthy approach, I believe it is one that will take at least a couple of decades to bear fruit. The quantity and diversity of low level neural circuitry that needs to be deciphered and modeled is huge. On the other hand, there is vast neuroscience knowledge today from which we can be inspired and attempt to design systems that mimic the concepts empowering the mammal brain. Such biologically-inspired work is picking up fast and can be seen in areas such as reinforcement learning and deeply-layered machine learning. Moreover, one can argue that while the mammal brain is our only instantiation of an intelligent system, it is most likely not an optimal (or the only possible) one. Thus, I believe being inspired by the brain, rather than trying to accurately reverse engineer the brain, is a more pragmatic path toward achieving AGI in the foreseeable future.

Ben

Another important, related conceptual question has to do with the relationship between mind and body in AGI.

In some senses, human intelligence is obviously closely tied to human embodiment – a lot of the brain deals with perception and action, and it seems that young children spend a fair bit of their time getting better at perceiving and acting. This brings up the question of how much sense it makes to pursue AI systems that are supposed to display roughly human-like cognition but aren't connected with roughly humanlike bodies. And then, if you do believe that giving an AI a human-like body is important, you run up against the related question of just *how* human-like an AI body needs to be, in order to serve as an appropriate vessel for a roughly human-like mind.

Itamar

What makes AGI truly challenging is its scalability properties. Machine learning has delivered numerous self-learning algorithms that exhibit impressive properties and results when applied to small scale problems. The real-world, however, is extremely rich and complex. The spatiotemporal patterns that we as human are continuously exposed to, and that help us understand the world with which we interact, are a big part of what makes achieving AGI a colossal challenge. To that end, while it is very possible to design, implement and evaluate AGI systems in simulated or virtual environments, the key issue is that of scalability and complexity. A physical body, facilitating physical interaction with the real-world, inherently offers reach stimuli from which the AGI system can learn and evolve. Such richness remains a challenge to be attained in simulation environments.

Ben

OK, so if we do interface our AGI systems with the physical world in a high-bandwidth way, such as via robotics – as you suggest – then we run up against some deep conceptual and technical questions, to do with the artificial mind's mode of interpreting the world.

This brings us to a particular technical and conceptual point in AGI design I've been wrestling with lately in my own work

Human minds deal with perceptual data and they also deal with abstract concepts. These two sorts of entities seem very different, on the surface – perceptual data tends to be effectively modeled in terms of large sets of floating-point vectors with intrinsic geometric structure; whereas abstract concepts tend to be well modeled in symbolic terms, e.g. as semantic networks or uncertain logic formulas or sets of various sorts, etc. So, my question is, how do you think the bridging between perceptual and conceptual knowledge works in the human mind/brain, and how do you foresee making it work in an AGI system? Note that the bridging must go both ways – not only must percepts be used to seed concepts, but concepts must also be used to seed percepts, to support capabilities like visual imagination.

Itamar

I believe that conceptually the brain performs two primary functions: situation inference and mapping of situations to actions. The first is governed by a perception engine which infers the state of the world from a sequence of observations, while the second maps this inferred state to desired actions. These actions are taken so as to maximize some reward driven construct. Both of these subsystems are hierarchical in nature and embed layers of abstraction that pertain to both perceptional concepts and actuation driven concepts. There is, indeed, continuous interdependency between those two subsystems and the information they represent, however the hierarchy of abstraction and its resulting representations are an inherent property of the architecture. My feeling is that the theory behind the components that would realize such architectures is accessible today.

Ben

Hmmm... You say "I believe that conceptually the brain performs two primary functions: Situation inference and mapping of situations to actions."

Now, it's moderately clear to me how most of what, say, a rabbit or a dog does can be boiled down to these two functions. It's less clear how to boil down, say, writing a sonnet or proving a mathematical theorem to these two functions. Any comments on how this reduction can be performed?

I suppose this is basically a matter of probing your suggested solution to the good old symbolic/connectionist dichotomy. Your AI approach is pretty much connectionist – the knowledge represented is learned, self-organized and distributed, and there's nothing like a built-in logic or symbolic representation system. So apparently you somehow expect these linguistic and symbolic functions to emerge from the connectionist network of components in your system, as a result of situation inference and situation-action mapping... And this is not a totally absurd expectation since a similar emergence apparently occurs in the human brain (which evolved from brains of a level of complexity similar to that of dogs and rabbits, and uses mostly the same mechanisms).

Itamar

I think you hit the nail on the head: Situation inference is a vague term that in practice involves complex, hierarchical and multi-scale (in time and space) abstraction and information representations. This includes the symbolic representations that are needed for a true AGI system. The "logic" part is a bit more complicated, as it resides on the intersection between the two primary subsystems (i.e. inference and control). The control part invokes representations that project to the inference engine, which in turn reflects back to the control subsystem as the latter eventually generates actuation commands. In other words, logical inference (as well as many "strategic" thinking capabilities) should emerge from the interoperability between the two primary subsystems. I hope that makes some sense!

Ben

I understand the hypothesis, but as you know I'm unsure if it can be made to work out without an awful lot of special tweaking of the system to get that emergence to happen. That's why I've taken a slightly different approach in my own AGI work, integrating an explicitly symbolic component with a connectionist "deep learning component" (which may end up being a version of your DeSTIN system, as we've been discussing!). But I certainly think yours is a very a worthwhile research direction and I'm curious to see how it comes out.

So anyway, speaking about that, how is your work going these days? What are the main obstacles you currently face? Do you have the sense that, if you continue with your present research at the present pace, your work will lead to human-level AGI within, say, a 10 or 20 year timeframe?

Itamar

In addition to my work at the University of Tennessee, I'm affiliated with a startup company out of the Silicon Valley called Binatix. Binatix aims to develop broadly applicable AGI technology. Currently, the work focuses on a unique deep machine learning based perception engine, with some very promising results to be made public soon. Down the road, the plan is to integrate a decision making subsystem as part of the work toward achieving AGI. My personal hope is that, by following this R&D direction, human-level AGI can be demonstrated within a decade.

Ben

You seem to feel you have a pretty solid understanding of what needs to be done to get to human-level AGI along a deep-learning path. So I'm curious to probe the limits of your knowledge! What would you say are the areas of AGI and cognitive science, relevant to your own AGI work, that you feel you understand *least* well and would like to understand better?

Itamar

I think that cognitive processes which take place during sleep sessions are critical to learning capabilities, and are poorly understood. The key question is whether sleep (or some analogous state) is a true prerequisite for machine intelligence, or will there be a way to achieve the same results attained via sleep by other means? This is something I'm working on at present.

Ben

I remember back in the 1990s when I was a cognitive science professor in Perth (in Western Australia), my friend George Christos there wrote a book called Memory and Dreams: The Creative Human Mind [8], about dreaming and its cognitive importance, and its analogues in attractor neural networks and

---

[8] http://www.amazon.com/Memory-Dreams-Creative-Human-Mind/dp/0813531306

other AI systems. He was playing around a lot with neural nets that (in a certain sense) dreamed. Since then there's been a lot of evidence that various phases of sleep (including but not only dream sleep) are important for memory consolidation in the human brain. In OpenCog it's pretty clear that memory consolidation can be done without a sleep phase. But yeah, I can see in your deep learning architecture, which is more brain-like than OpenCog, it's more of a subtle question – more like with the attractor neural nets George was playing with.

Itamar

Also, I think the question of "measuring" AGI remains an unsolved one, for which there needs to be more effort on the part of the community. If we hope to be on the path to AGI, we need to be able to measure progress achieved along such path. That requires benchmarking tools, which are currently not there. It would be interesting to hear what other AGI researchers think about AGI benchmarking, etc.

Ben

Hmmm... Well – taking off my interviewer hat and putting on my AGI researcher hat for a moment – I can tell you what I think. Honestly, my current feeling is that benchmarking is fairly irrelevant to making actual research progress toward AGI. I think if you're honest with yourself as a researcher, you can tell if you're making worthwhile progress or not via careful qualitative assessments, without needing formal rigorous benchmarks. And I'm quite wary of the way that formulation and broad acceptance of benchmark tests tends to lead to a "bakeoff mentality"[9] where researchers focus too much on beating each other on test criteria rather than on making real research progress.

On the other hand, I can certainly see that benchmarking is useful for impressing funding sources and investors and so forth,

---

[9] http://cll.stanford.edu/~langley/papers/eval.permis04.pdf

because it gives a way for them to feel like concrete progress is being made, well before the end goal is achieved.

As you know we discussed these issues at the AGI Roadmap Workshop we co-organized in 2009... I think we made some good progress toward understanding and articulating the issues involved with assessing progress toward AGI – but we also came up against the sheer difficulty of rigorously measuring progress toward AGI. In particular it seemed to be very difficult to come up with tests that would usefully measure partial progress toward AGI, in a way that couldn't be "gamed" by narrow-AI systems engineered for the tests...

It's been a while since that workshop now – I wonder if you have any new thoughts on what kinds of benchmark tests would be useful in the context of your own AGI research program?

Itamar
Unfortunately, I don't. Like you, I think that one would know if he/she were truly far along the path to introducing true AGI. It's a bit like being pregnant; you can't be "slightly" pregnant :-)

Ben
Yeah. and I tend to think that, at a certain point, some AGI researcher – maybe one of us; or maybe the two of us working together, who knows! – is going to produce a demonstration of AGI functionality that will be sufficiently compelling to make a significant percentage of the world feel the same way. I think of that as an "AGI Sputnik" moment – when everyone can look at what you've done and qualitatively feel the promise. That will mean more than progress along any quantitative benchmarks, I feel. And after the AGI Sputnik moment happens, things are going to progress tremendously fast – both in terms of funding and brainpower focused on AGI, and (correlatedly) in terms of R&D progress.

Pei Wang: What Do You Mean by "AI"?

Interview first published January 2011

Pei Wang is one of the most important AGI researchers in my own personal history. Pei and I disagree on many things regarding AGI design and philosophy, but he has also inspired me in many ways. He was one of the first employees I hired for my late-1990s AI startup Webmind Inc. (initially named Intelligenesis), and during the 3 years he worked for me we shared an endless series of fascinating discussions on the various practical and conceptual aspects of (what would later come to be called) AGI. His work on the NARS Non-Axiomatic Reasoning System was one of the inspirations for my own work on the PLN Probabilistic Logic Networks system, which is a key part of my OpenCog integrative AGI architecture. In 2009 he provided invaluable help with the AGI Summer School I organized in Xiamen, China. The interview I did with Pei, that's reported here, is only a tiny summary and snippet from our decades-long, multi-dimensional dialogue.

After first getting his Computer Science education in China, Pei came to the US to complete his PhD with famed writer and cognitive scientist Douglas Hofstadter. Since that time he's held a variety of university and industry positions, including a stint working for me in the dot-com era as I mentioned above. Currently he's a computer science faculty at Temple University, and pushing ahead his approach to AGI based on his NARS[10] system (Non-Axiomatic Reasoning System).

---

[10] http://sites.google.com/site/narswang/home/nars-introduction

Those with a technical bent may enjoy Pei's 2006 book *Rigid Flexibility: The Logic of Intelligence*[11] Or for a briefer summary, see Pei's Introduction to NARS[12] or his talk from the 2009 AGI Summer School[13]

Ben

I'll start out with a more "political" sort of question, and then move on to the technical stuff afterwards.

One of the big complaints one always hears among AGI researchers is that there's not enough funding in the field. And indeed, if you look at it comparatively, the amount of effort currently going into the creation of powerful AGI right now is really rather low, relative to the magnitude of benefit that would be obtained from even modest success, and given the rapid recent progress in supporting areas like computer hardware, computer science algorithms and cognitive science. To what do you attribute this fact?

I note that we spend a lot of money, as a society, on other speculative science and engineering areas, such as particle physics and genomics – so the speculative nature of AGI can't be the reason, in itself, right?

Pei

We addressed this issue in the Introduction[14] we wrote together for the book *Advances in Artificial General Intelligence*, back in 2006, right? We listed a long series of common objections given

---

[11] http://www.amazon.com/Rigid-Flexibility-Logic-Intelligence-ebook/dp/B001BTESQQ

[12] http://sites.google.com/site/narswang/home/nars-introduction

[13] http://agi-school.org/2009/dr-pei-wang-a-logical-model-of-intelligence-1-of-3

[14] http://goertzel.org/agiri06/%5B1%5D%20Introduction_Nov15_PW.pdf

by skeptics about AGI research, and argued why none of them hold water.

Ben

Right, the list was basically like this:

- AGI is scientifically impossible
- There's no such thing as "general intelligence"
- General-purpose systems are not as good as special-purpose ones
- AGI is already included in current AI
- It's too early to work on AGI
- AGI is nothing but hype; there's no science there
- AGI research is not fruitful; it's a waste of effort
- AGI is dangerous; even if you succeed scientifically you may just destroy the world

I guess the hardest one of these to argue against is the "too early" objection. The only way we can convincingly prove to a skeptic that it's not too early, is to actually succeed at building an advanced AGI. There's so much skepticism about AGI built up, due to the failures of earlier generations of AI researchers, that no amount of theoretical argumentation is going to convince the skeptics. Even though now we have a much better understanding of the problem, much better computers, much better neuroscience data, much better algorithms, and so forth.

Still, I think the situation has improved a fair bit – after all, we now have AGI conferences every year, a journal dedicated to this topic[15], and there are at least special sessions on human-level AI and so forth within the mainstream AI conferences. More so than 10 or even 5 years ago, you can admit you work on AGI

---

[15] http://journal.agi-network.org/

and not get laughed at. But still, the funding isn't really there for AGI research, the way it is for fashionable types of narrow AI.

What do you think – do you think the situation has improved since we wrote that chapter in 2006?

Pei

It has definitely improved, though not too much.

Within the fields of artificial intelligence and cognitive science, I think the major reason for the lack of effort in the direction of AGI right now is the well-known difficulty of the problem. I suppose that's much the same as the "too early" objection that you mention. To most researchers, it seems very hard, if not impossible, to develop a model of intelligence as a whole. Since there is no basic consensus on the specific goal, theoretical foundation, and development methodology of AGI research, it is even hard to set up milestones to evaluate partial success. This is a meta-level problem not faced by the other areas you mentioned – physics and biology and so forth.

Furthermore, in those other areas there are usually few alternative research paths – whereas in artificial intelligence and cognitive science, researchers can easily find many more manageable tasks to work on, by focusing on *partial problems*, and still get recognition and rewards. Actually the community has been encouraging this approach by defining "intelligence" as a loose union of various "cognitive functions", rather than a unified phenomenon.

Ben

One approach that has been proposed for surmounting the issues facing AGI research, is to draw on our (current or future) knowledge of the brain. I'm curious to probe into your views on this a bit.

Regarding the relationship between neuroscience and AGI, a number of possibilities exist. For instance, one could propose

a) to initially approach AGI via making detailed brain simulations, and then study these simulated human brains to learn the principles of general intelligence, and create less humanlike AGIs after that based on these principles; or

b) to thoroughly understand the principles of human intelligence via studying the brain, and then use these principles to craft AGI systems with a general but not necessarily detailed similar to the brain; or

c) to create AGI systems with only partial resemblance to the human brain/mind, based on integrating our current partial knowledge from neuroscience with knowledge from other areas like psychology, computer science and philosophy; or

d) to create AGI systems based on other disciplines without paying significant mind to neuroscience data.

I wonder, which of these four approaches do you find the most promising, and why?

Pei

My approach is roughly between the above D and C. Though I have gotten inspirations from neuroscience on many topics, I do not think to build a detailed model of neural system is the best way to study intelligence.

I said a lot about this issue in the paper "What Do You Mean by AI?[16]" in the proceedings of the AGI-08 conference. So now I'll just briefly repeat what I said there.

The best known object showing "intelligence" is undoubtedly the human brain, so AI must be "brain-like" in some sense. However, like any object or process, the human brain can be studied and modeled at multiple levels of description, each with its

---

[16] http://sites.google.com/site/narswang/publications/wang.AI_Definitions.pdf?attredirects=0

vocabulary, which specifies its granularity, scope, visible phenomena, and so on. Usually, for a given system, at a lower level descriptions say more on its internal structure, while at a higher level descriptions say more on its overall regularity and outside interaction. No level is "more scientific" or "closer to reality" than the others, so that all the other levels can be reduced into it or summarized by it. As scientific research, study on any level can produce valuable theoretical and practical results.

However, what we call "intelligence" in everyday language is more directly related to a higher level description than the level provided by neuroscience. Therefore, though neuroscientific study of the brain can gradually provide more and more details about the mechanism that supports human intelligence, it is not the most direct approach toward the study of intelligence, because its concepts often focus on human-specific details, which may be neither necessary nor possible to be realized in machine intelligence.

Many arguments supporting the neuroscientific approach toward AGI are based on the implicit assumption that the neuro-level description is the "true" or "fundamental" one, and all higher-level descriptions are its approximations. This assumption is wrong for two reasons: (1) Such a strong reductionist position has been challenged philosophically, and it is practically impossible, just like no one really wants to design an operating system as a string of binary code, even though "in principle" it is possible. (2) Even according to such a strong reductionist position, the "neural level" is not the lowest level, since it is not hard to argue for the contribution to intelligence from the non-neuron cells, or the non-cell parts or processes in the human body.

Ben

Another important conceptual question has to do with the relationship between mind and body in AGI.

In some senses, human intelligence is obviously closely tied to human embodiment – a lot of the brain deals with perception and action, and it seems that young children spend a fair bit of their time getting better at perceiving and acting. This brings up the question of how much sense it makes to pursue AI systems that are supposed to display roughly human-like cognition but aren't connected with roughly humanlike bodies. And then, if you do believe that giving an AI a human-like body is important, you run up against the related question of just *how* human-like an AI body needs to be, in order to serve as an appropriate vessel for a roughly human-like mind.

(By a "roughly human-like mind" I don't mean a precisely simulated digital human, but rather a system that implements the core principles of human intelligence, using structures and processes with a reasonable conceptual correspondence to those used by the human mind/brain.)

Pei

Once again, it depends on "What do you mean by AI?" Even the above innocent-looking requirement of "a reasonable conceptual correspondence to those used by the human mind/brain" may be interpreted very differently in this context. If it means "to respond like a human to every stimulus", as suggested by using the Turing Test to evaluate intelligence, then the system not only needs a (simulated) human body, but also a (simulated or not) human experience. However, as I argued in the AGI-08 paper, if "intelligence" is defined on a more abstract level, as a human-like experience-behavior relation, then a human-like body won't be required. For example, an AGI system does not need to feel "hungry" in the usual sense of the word (that requires a simulated human body), but it may need to manage its own energy repository (that does not require a simulated human body). This difference will surely lead to difference in behaviors, and whether such a system is still "intelligent" depends on how the "human-like" is interpreted.

Ben

Hmmm... I understand; but this doesn't quite fully answer the question I was trying to ask.

My point was more like this: Some researchers argue that human thinking is fundamentally based on a deep and complex network of analogies and other relationships to human embodied experience. They argue that our abstract thinking is heavily guided by a sort of visuomotor imagination, for example. That our reasoning even about abstract things like mathematics or love is based on analogies to what we see and do with our bodies. If that's the case, then an AGI without a humanlike body might not be able to engage in a humanlike pattern of thinking.

Pei

The content of human thinking depends on human embodied experience, but the mechanism of human thinking doesn't (at least not necessarily so).

If a robot has no vision, but has advanced ultrasonic sensation, then, when the system has AGI, it will develop its own concepts based on its own experience. It won't fully understand human concepts, but we cannot fully understand its, neither. Such a system can develop its "math" and other abstract notions, which may be partially overlap with ours, though not completely. According to my definition, such a system can be as "intelligent" as human, since its experience-behavior relationship is similar to ours (though not the experience, or behavior, separately). By "abstract", I mean the meta-level mechanism and processes, not the abstract part of its object-level content.

Ben

Yes, but... I'm not sure it's possible to so strictly draw a line between content and mechanism.

Pei

Of course, it is a matter of degree, but to a large extent the distinction can be made. On a technical level, this is why I prefer the "reasoning system" framework – here "object language" and "meta language" are clearly separated.

Ben

The mechanism of human thinking is certainly independent of the specific content of the human world, but it may be dependent in various ways on the "statistics" (for lack of a better single word) of the everyday human world. For instance, the everyday human world is full of hierarchical structures; and it's full of solid objects that interact in a way that lets them maintain their independence (very different than the world of, say, a dolphin or an intelligent gas cloud on Jupiter – I wrote an article[17] last year speculating on the properties of intelligences adapted for fluid environments). And the brain's cognitive mechanisms may be heavily adapted to various properties like this, that characterize the everyday human world. So one line of thinking would be: If some AGI's environment lacks the same high-level properties as the everyday human world, then the cognitive mechanisms that drive human-level intelligence may not be appropriate for that AGI.

Pei

Can you imagine an intelligence, either in the AI context or the *Alien Intelligence* context, to have sensors and motors very different from ours? If the answer is yes, then the *content* of their mind will surely be different from ours. However, we still consider them as "intelligent", because they can also adapt to their environment, solving their problems, etc., and their adaptation and problem-solving mechanisms should be similar to ours (at least I haven't seen why that is not the case) – all intelligent systems need to summarize their experience, and use the pattern observed to predict the future.

---

[17] http://journalofcosmology.com/SearchForLife115.html

I agree with you that in different environments, the most important "patterns" may be different, which will in turn favor different mechanisms. It is possible. However, the opposite is equally possible (and also interesting to me), that is, no matter in which environment, the major mechanism for adaptation, recognition, problems-solving, etc., is basically the same, and its variations can be captured as different parameter settings. This mechanism is what "intelligence" means to me, not the concrete beliefs, concepts, and skills of the system, which depend on the concrete body and environment.

Ben

On the other hand, a virtual world like Second Life also has a sort of hierarchical structure (though not as rich or as deeply nested as that of the everyday human physical world), and also has solid objects – so for the particular two high-level properties I mentioned above, it would potentially serve OK.

Pei

Sure, though there is a difference: the hierarchical structures in the natural world is largely the result of our mental reconstruction from our experience, so there is no "correct answer", while the current virtual worlds are not that rich (which may change in the future, of course).

Ben

Also, one could argue that some cognitive mechanisms only work with a very rich body of data to draw on, whereas with a data-poor environment they'll just give nonsense results. For instance (and I say this knowing you don't consider probability central to AGI, but it's just an example), some probabilistic methods require a large number of cases in order to meaningfully estimate distributions.... In that case, such a cognitive mechanism might not work well for an AI operating in Second Life, because of the relative poverty of the data available...

Pei

In principle, I agree with the above statement, but I don't think it mean that we cannot distinguish mechanism from content.

Ben

So I am I correct in understanding that, in your view, the same basic cognitive mechanisms are at the core of AGI no matter what the high-level properties of the environment (not just no matter what is the specific content of the environment)?

Pei

Yes, though I don't mean that they are the best solution to all practical problems. In certain environments (as you mentioned above), some less intelligent mechanisms may work better.

Ben

Hmmm... Or, alternately, do you think that the cognitive mechanisms of general intelligence are tied to *some* high-level properties of the environment, but that these properties are *so* high-level that any environment one gives an AGI system is likely to fulfill them?

Pei

I'd say that intelligence works in "most interesting environments". If the environment is constant, then the traditional computational models are better than intelligent ones; on the other extreme, if the environment is purely arbitrary, and no pattern can be recognized in meaningful time, intelligence is hopeless. However, since I'm not looking for a mechanism that is optimal in all environments, it is not an issue to me.

Ben

OK, well I don't think we can go much deeper in that direction without totally losing our audience! So I'll veer back toward the "politics" side of things again for a bit... Back to the nasty business of research funding!

As we both know, "narrow AI" research (focusing on AI programs that solve very specific tasks and don't do anything else, lacking the broad power to generalize) gets a lot more attention and funding than AGI these days. And in a sense this is understandable, since some narrow AI applications are delivering current value to many people (e.g. Internet search, financial and military applications, etc.). Some people believe that AGI will eventually be achieved via incremental improvement of narrow-AI technology – i.e. that narrow AI can gradually become better and better by becoming broader and broader, until eventually it's AGI. What are your views on this?

Pei

"Narrow AI" and "AGI" are different problems, to a large extent, because they have different goals, methods, evaluation criteria, application, etc., even though they are related here or there. I don't think AGI can be achieved by integrating the existing AI results, though these tools will surely be useful for AGI.

Ben

Yeah, as you know, I agree with you on that. Narrow AI has yielded some results and even some software that can be used in building AGI applications, but the core of an AGI system has got to be explicitly AGI-focused AGI can't be cobbled together from narrow AI applications. I suppose this relates to the overall the need to ground AGI work in a well-thought-out philosophy of intelligence.

Pei

Yes. One problem slowing down progress toward AGI, I think, has been a neglect among AI researchers of a related discipline: philosophy. Most major mistakes in the AGI field come from improper philosophical assumptions, which are often implicitly held. Though there is a huge literature on the philosophical problems in AI, most of the discussions there fail to touch the most significant issues in the area.

Ben

So let's dig a little into the philosophy of mind and AI, then. There's one particular technical and conceptual point in AGI design I've been wrestling with lately in my own work, so it will be good to get your feedback.

Human minds deal with perceptual data and they also deal with abstract concepts. These two sorts of entities seem very different, on the surface – perceptual data tends to be effectively modeled in terms of large sets of floating-point vectors with intrinsic geometric structure; whereas abstract concepts tend to be well modeled in symbolic terms, e.g. as semantic networks or uncertain logic formulas or sets of various sorts, etc. So, my question is, how do you think the bridging between perceptual and conceptual knowledge works in the human mind/brain, and how do you foresee making it work in an AGI system? Note that the bridging must go both ways – not only must percepts be used to seed concepts, but concepts must also be used to seed percepts, to support capabilities like visual imagination.

Pei
I think I look at it a little differently than you do. To me, the difference between perceptual and conceptual knowledge is only "on surface", and the "vectors vs. symbols" distinction merely shows the choices made by the previous researchers. I believe we can find unified principles and mechanisms at both the perceptual level and the conceptual level, as a continuous and coherent "conceptual hierarchy". Here "conceptual" means "can be recognized, recalled, and manipulated as a unit within the system", so in this broad sense, various types of percepts and actions can all be taken as concepts. Similarly, perceptual and conceptual knowledge can be uniformly represented as specialization-generalization relations among concepts, so as to treat perception and cognition both as processes in which one concept is "used as" another in certain way.

According to this view the distinction between perceptual and conceptual knowledge still exists, but only because in the conceptual hierarchy certain concepts are closer to the sensors and actuators of the system, while some others are closer to the words in communication languages of the system. Their difference is relative, not absolute. They do not need to be handled by separate mechanisms (even with a bridge in between), but by a uniform mechanism (though with variants in details when it is applied to different parts of the system).

Ben

Certainly that's an elegant perspective, and will be great if it works out. As you know my approach is more heterogeneous – in OpenCog we use different mechanisms for perception and cognition, and then interface them together in a certain way.

To simplify a bit, it feels to me like you begin with cognition and then handle perception using mechanisms mainly developed for cognition. Whereas, say, Itamar Arel or Jeff Hawkins, in their AGI designs, begin with perception and then handle cognition using mechanisms mainly developed for perception. On the other hand, those of us with fundamentally integrative designs, like my OpenCog group or Stan Franklin's LIDA approach, start with different structures and algorithms for perception and cognition and then figure out how to make them work together. I tend to be open-minded and think any of these approaches could potentially be made to work, even though my preference currently is for the integrative approach.

So anyway, speaking of your own approach – how is your own AGI work on NARS going these days? What are the main obstacles you currently face? Do you have the sense that, if you continue with your present research at the present pace, your work will lead to human-level AGI within, say, a 10 or 20 year timeframe?

Pei

My project NARS has been going on according to my plan, though the progress is slower than I hoped, mainly due to the limit of resources.

What I'm working on right now is: Real-time temporal inference, emotion and feeling, self-monitoring and self-control.

If it continues at the current pace, the project, as currently planned, can be finished within 10 years, though whether the result will have "human-level AGI" depends on what that phrase means – to me, it will have.

Ben

Heh... Your tone of certainty surprises me a little. Do you really feel like you know for *sure* that it will have human-level general intelligence, rather than needing to determine this via experiment? Is this certainty because you are certain your theory of general intelligence is correct and sufficient for creating AGI, so that any AGI system created according to this theory will surely have human-level AGI

Pei

According to my theory, there is no absolute certainty on anything, including my own theory! What I mean is: according to my definition of "intelligence", I currently see no major remaining conceptual problem. Of course we still need experiments to resolve the relatively minor (though still quite complicated) remaining issues.

Ben

Indeed, I understand, and I feel the same way about my own theory and approach! So now I guess we should stop talking and get back to building our thinking machines. Thanks for taking the time to dialogue; I think the result has been quite interesting.

Joscha Bach:
Understanding the Mind

Interview first published May 2011

I have a lot of good things to say about Joscha Bach. He has a deep conceptual understanding of AGI and its grounding in cognitive science, philosophy of mind and neuroscience; a software system embodying a subset of his ideas and doing interesting and useful things; and a concrete plan for what needs to be learned to expand his system and his associated theory gradually toward human-level intelligence. What's more, I actually think his approach might work, if extended and grown intelligently from its current state. There aren't many AGI researchers I can say all that about.

Joscha got a strong start in AGI via launching the MicroPsi project, a cognitive architecture based on an elaboration of Dietrich Dörner's Psi model of motivated cognition. To test and illustrate the theory, Joscha used MicroPsi to control a population of simulated agents in a simple virtual world, as they interacted with each other and sought food and water and other resources in order to conduct their virtual lives. He also experimented with other AGI-relevant aspects of AI such as vision processing and robotics.

In 2005 Joscha left academia to co-found txtr, a company marketing e-reader software (and, briefly, e-reader hardware as well). But he continued to find time for AGI thinking, publishing his excellent book *Principles of Synthetic Intelligence* in 2009, summarizing his early work on MicroPsi and taking a few additional steps. While he found entrepreneurship an invigorating learning experience, and txtr is proceeding successfully, in 2011 he returned to academia, focusing back on AGI research full-time via a position at Humboldt University in

Berlin. Now he is building a new version of MicroPsi, aiming at both research and commercial applications.

I've found some of Joscha's scientific ideas useful in my own AGI work, and the OpenCog AGI project that I co-direct has recently adopted some MicroPsi-based notions in its treatment of goals, actions and emotions. This indicates that we see eye to eye on many aspects of the AGI problem – and indeed, one amusing thing I found when editing this dialogue, was that I occasionally got confused about which parts were said by Joscha, and which parts were said by me! That very rarely happens to me when editing interviews or dialogues, and indicates that some aspects of our thinking on AGI are very closely aligned.

However, Joscha and I also disagree on many details regarding the best approach to AGI, which sometimes leads to animated discussions – for instance during his 2011 visit to the OpenCog game-AI project in Hong Kong. I'm afraid the dialogue between us presented below doesn't quite capture the depth or the energy of our face-to-face debates. A series of words on a screen or printed page can't quite capture the spirit of, say, our heated debate about the merits of multimodal story understanding as an AGI test domain, conducted while walking past the statue of Bruce Lee in Kowloon by the Hong Kong harbor. Also, many of our core points of dissension are too technical to be compactly discussed for a general audience (for instance, just to give the flavor of the sorts of issues involved: his preference for a more strictly "subsymbolic" network of concept nodes, versus OpenCog's more flexible concept node network that mixes subsymbolic representations with more explicit rule type representations). But even so, I think this dialogue captures a few of the many interesting aspects of our AGI discussions.

Ben
You've said that your core motivation as an AGI researcher is to understand intelligence, to understand the mind. Can you explain

why you think AGI is a good way to go about seeking this kind of understanding?

Joscha

Intelligence is probably not best understood as a collection of very particular capabilities and traits of test subjects (like in experimental psychology), nor as an emanation of a particular conglomerate of biological cells, their chemistry or their metabolic rates. Why not? Imagine you were to study the principles of flying: should you categorize birds, or examine their muscle cells? While both details might be helpful along the way, they are probably neither necessary nor sufficient for your goal. Instead, the most productive approaches have been the formulation of a theory of aerodynamics (the correct functional level for flight) along with the building of working models. This is exactly what AGI proposes: deriving, and simulating theories of intelligent information processing, including perception, motivation, memory, learning, imagination, language personhood and conscious reflection.

Ben

What jumps out at me here is that aerodynamics is a unified theory with a few core principles, and then a bunch of specialized subtheories derived from these principles plus various special assumptions. Whereas, in the case of AGI you give a sort of laundry list of areas for theories to address: "perception, motivation, memory, learning, imagination, language, personhood and conscious reflection...".

The entries on this list seem to correspond better to subtheories of aerodynamics covering particular cases or aspects of fluid flow and flight, rather than to the core principles of aerodynamics. So this leads up to a question: do you think there will emerge some core set of mathematical/conceptual principles of intelligence, with the generality and power of, say, the Navier-Stokes equations that underlie fluid dynamics (and hence aerodynamics)? Or is the underlying theory of intelligence going to be more messy and laundry-list-like? If the former, what sorts of ideas do you see as most likely to underlie this potential unifying underlying theory of intelligence?

Joscha

There are two main directions of approach towards AGI: One is constructive. We identify each sub-problem, perform a conceptual analysis, derive tasks, specifications and benchmarks, and get to engineering. Let us call the other one 'emergentist'. The idea is to specify a set of general dynamic information processing principles and constraints, and arrange them in a relatively simple top-level layout. After a period of sensory-motor learning, micro and macro structure of an artificial mind/brain are going to emerge.

There is some correspondence between these approaches and the divide between 'classical', symbol processing AI, and the currently fashionable 'embodied' AI, which eschews symbols, language, abstract decision making, planning and so on. If we understand symbol processing as logic oriented systems, then we are probably looking at the wrong paradigm. I have little hope for emergent embodied AI either. It is true that this is the way humans work: our genes define both a set of dynamic principles (for neural learning, generation of maps and clusters, reorganization according to new stimuli etc.) and the overall layout of our nervous system. But even if we manage to get the principles right in our models, and we recreate the major known features of the cortical architecture (cell types, neurotransmitters, columns, layers, areas, main pathways and so on) – and this is a big if we might not end up with something intelligent, in the sense that it is capable of learning a natural language or apply itself to an engineering problem. Nature has brought forth a huge number of these sensorimotor emergentist designs, and none, with the single exception of homo sapiens, is generally intelligent.

I suspect that we need a toolbox of the principles and constraints I just mentioned, so we can endow our system with cognitive information processing. This toolbox might be akin to the principles of aerodynamics, but I guess a little more complex. But just as the theory of aerodynamics does not yet imply the design of a particular helicopter, this alone is not enough. We will need to apply this toolbox to engineer the particular set of cognitive behaviors that is necessary for AGI.

By the way: "Cognitive AI" might be a very good label for what we want to do. It is not about Symbolic or Embodied AI (most embodied critters are not smart). The notion of Generality (as in 'Artificial General Intelligence') might be too difficult to define to be truly useful. (Does anybody remember the General Problem Solver?) But 'cognition' might hit the right set of problems: what is it that enables a system to think, to perceive, to learn? What representations are suitable for cognition? What motives are necessary to direct it? What kind of environmental interaction is involved? These are the questions we need to ask.

Ben

You've posed a lot of interesting ideas there, which we could dissect in detail. But one thing that jumps out at me is your perspective on embodiment. You seem to imply it's not so central to the quest for understanding intelligence. Yet your own MicroPsi project did focus on embodiment of a sort – using your AI architecture to control virtual characters in a simple simulation world. Which seems to suggest you find some level of embodiment valuable.

Joscha

Some kind of embodiment is valuable for testing and teaching a cognitive architecture. However, I think a human-like AI could probably live in the body of a kraken, or in World of Warcraft, as long as the internal architecture – the motivational and representational mechanisms and the structure of cognitive processes are similar to the one of humans, and a sufficient (physical or virtual) environment is provided.

Ben

That's an interesting direction to me. Care to share some more detailed thoughts on what constitutes a "sufficient" environment for human-like AGI? The relative paucity of stimuli in both World of Warcraft and the typical robot lab (as compared to everyday human environments) is on my mind here.

Joscha

You're correct regarding robot labs. A lot of AI scientists defend their robot grants with the need to go beyond the paucity of simulations. And then they go ahead and put their robot in a black and white labyrinth, or in the simplistic green-white-orange environment of robotic soccer. Current robotic environments are not about complex social interaction or applied real-world problem solving. Instead, they are usually about limited locomotion, combined with second-rate image recognition. But I guess your question does not address the problems of using robots instead of simulations. So, what's necessary and sufficient, if robotics are neither?

Human intelligence is the answer to a complex and badly specified set of problems, such as: How can I find food, shelter, intellectual growth and social success? Tackling these problems over the course of one or two decades leads to a bunch of encodings in our nervous systems, along with a set of operations to manipulate these encodings, so we can infer, reflect, anticipate and plan. The encodings in our brain represent the order that our cognition managed to discern in the patterns thrown at it by the environment.

Epistemologically, it does not matter how we come by these encodings. An AI might get them by (A) proper perceptual and developmental learning, (B) by offline inference from a large corpus of environmental patterns, (C) by programming them manually, or (D) by copying them from another intelligent system with a similar architecture. I do not see how we can do (D) in the near future, or how (C) is tractable, so let us look at the other options.

If you go for (A), you will need an environment that provides situatedness and supplies hierarchical, compositional objects, which the system can rearrange in interesting and useful ways. It should provide obstacles and ways to overcome them by applying problem solving capabilities. To make the obstacles count as such, they should block the way to needs (such as nutrition, shelter, social success, intellectual growth), and it probably helps if there are multiple problem dimensions, so that optimization is non-trivial. Free compositional thought seems to require the structuring provided by a natural language, so the environment needs language teachers and stuff to talk about. Also, we want a person-like and adult AI, the environment needs to provide social interaction and even a formal education. I do not think that any current robotic environment is even remotely sufficient for this.

Also, (B) might go a long way: statistical methods, applied on Youtube and the textual World Wide Web, might be able to infer much the same representations as humans have. We will still need to find the right representational framework, and the right operations and inference methods, but perhaps we won't need an environment at all.

That said, we should consider research trajectories. I believe that developmental learning in a complex environment could provide an excellent, competitive benchmark, and an exciting research task as well. If we move in that direction, we will see a continuous competition between (A), (B) and (C), and combinations thereof.

Ben

You mention perceptual learning, but the virtual world you used for MicroPsi was somewhat simplistic in terms of the perceptual challenges it offered. But I know you've also done some work on computer vision. Also, as you know, our friend Itamar Arel is one of many AGI researchers who believes modeling complex environments – unlike those in current simulated worlds or robot labs – is critical for human-like intelligence, perhaps even the most critical thing. What do you think? How important is perceptual modeling to AGI, versus abstract conceptual thinking? Or does it even make sense to consider the two separately?

Joscha

I think there is probably no fundamental difference between percepts and concepts; rather, percepts are something like conceptual primitives.

Ben

Hmmm... But do you think everything in the mind is built up from percepts, then? Do you think there are other conceptual primitives besides percepts? Mathematical primitives, or primitive operations of mental self-reflection, for example?

Joscha

At the moment, I do not think that we would need other representational primitives. I guess that the initial building blocks of mathematics are inferred as generalizations from perceptual learning. I do not know the set of primitive operations for cognitive processing, though.

Ben

Fascinating topic indeed... But while it's tempting to bounce my ideas about primitive cognitive operations off you, this interview probably isn't the best context for that! So now I'd like to shift our chat in a more practical direction. As we've discussed before, while our philosophies of AGI are broadly compatible, I'm a bit more optimistic than you are. Assuming my OpenCog design is roughly adequate for human-level AGI (which, if so, I wouldn't take as imply your design or others are inadequate... I think there may be many possible routes), I estimate there's maybe 50-70 expert man-years of work between here and human-level AGI. I accept that could be wrong, but I think it's the right order of magnitude.

Joscha

I believe that the creation of an AGI exceeds the scope of 50 man-years by a wide margin, probably by more than one order of magnitude.

Ben

I'd love to hear more of the reasoning underlying your estimate...

Joscha

You are talking about something in the order of 10 million dollars. I have an idea how far a startup company, or a university project, can go for that much money, and your estimate strikes me as extremely optimistic. I remember when we were discussing that a budget like the one for the Avatar movie would really make a difference for AGI. 10 million – that's roughly twice the budget for Kurzweil's Singularity movie, or about as much as Sex and the City II spent on their wardrobe. Or, to transfer it to software development: 10 million dollars is an average budget for the development of a PC or PlayStation game. Even the environment to put the fledgling AI into is likely to cost more than that.

Ben

Certainly, a government agency or a large company (or a bureaucratic small company) could not create an AGI with 50 man-years of work, or with 10 million dollars. And a startup in Silicon Valley or a major US university has huge cost overheads. But a small group of the right dedicated people with the right ideas, in an inexpensive location, can do a lot with fairly minimal resources.

Joscha

Yes, that's true, of course. However, in most cases where that happens, the small, dedicated team involved is solving a single hard problem. AGI is not concerned with a single enigmatic obstacle, but with at least several dozen problem fields. Each of these fields presents an infuriating number of puzzles.

Ben

Again this seems to come back to the "laundry list" versus "core unifying theory" dichotomy. What some AGI researchers would say is that, if you get the core unifying ideas of your AGI architecture right, then the puzzles in these various specialized areas will be dramatically simplified even if not altogether eliminated. I admit I have some sympathy for this perspective. I think the problem remains hard once you understand the core principles, but not quite as hard as you're saying...

Joscha

To get back to my slightly crude airplane metaphor: A theory of aerodynamics will drastically cut costs and time in the design of an airplane. It is probably even a prerequisite. And yet, it won't solve the problems of constructing the airplane itself: what should the ailerons look like, or can we do without them? Where do we put the wheels, and how do we brake them? How do we distribute the payload, and which material is suitable for stiffening the wings, and so on.

A lot of the AGI problems are similar, in that they do not represent general principles of neural information processing. I think that humans share these general principles with other mammals, and yet, none of these others are generally intelligent. The architecture of the human brain is not just the result of general principles. It has additional constraints, so that particular lesions to the brain often yield specific defects. Some of these defects can be compensated for, which is an argument for general principles at work--but in general, this is not true.

Each problem is likely to require its own solution – there seems to be no magic bullet that somehow kills them all.

Ben

Right, but what some would say is that a good unifying perspective can simplify these problems a lot, even if not automagically solving them all...

Joscha

Even worse, they all have to be treated as parts of an integral, whole design--for instance, a beautiful category learning paradigm is worthless if is not integrated with visual perception, action categories or goal structures.

Ben

Sure – but this "cognitive synergy" constraint may actually simplify the solution of the various subordinate problems, because often the solution to a problem in one area (vision, action, goals, etc.) may be to appropriately rely on work the mind is doing in other areas... So a good unifying perspective might simplify the problems of the individual areas of intelligence by appropriately and approximately reducing them to problems of how to fine-tune the interactions between that area and other areas.

Joscha

A lot of people can claim to have been there, done that. Unified cognitive architectures, like John Anderson's ACT and Soar, by Newell, Laird and Rosenbloom promised to have found the principles. Unfortunately, the little things did not fall into place just yet – each architecture just gave us a new way of looking at them.

The right toolbox of principles will help in solving many problems, and make some even disappear. But how do we get there? We already know that we are not going to get the unified principles right at the first attempt, because we (as a field) are already trying for decades. Our progress can only be measured by results, so we have to attack many of the 'laundry-list' problems. Our successes and failures will give us in turn indication on how we are doing on the general principles. Based on this knowledge, we improve the principles, but will then likely need to revisit the problems that we already considered solved.

We cannot really hope to be able to solve the sub-problems without the proper set of unified principles, and we will not find the principles without generalizing over the solution of the sub-problems. It is the classical chicken-and-egg problem. The answer to chicken-and-egg problems is usually called bootstrapping. In the beginning, we might resort to broad and shallow architectures – systems that can do a little bit of everything, but almost nothing completely right. But eventually, we have to add depth, and each bootstrapping step might span several dissertations, for instance, a complete visual learning paradigm.

This methodology is not very suitable for academic research, and does not work at all with industrial research. It may initially require us to build systems with a lower performance on the laundry-list items than already existing solutions. We will have to be willing to ditch nicely working (and publishable) paradigms whenever progress in the underlying theory requires it. We will have to rewrite stuff that worked well last year, and ask the same of our disciples, and we have to kick out impressive pieces of engineering if they are not integrated with the core.

Practically, all this is almost impossible to do right. And yet it has been fun and productive in the past, and I suspect it will be for some time to come.

Ben

Actually I agree with all that! I guess I just don't feel our current understanding of the principles of general intelligence is as inadequate as you do. But that would lead us into a deeper discussion of what we each respectively think the principles of intelligence are – which is too much for a dialogue like this, or at least deserves a whole other dialogue! In fact it occurs to me we've never really had that conversation face-to-face or via private email, at least not in a completely direct way. Perhaps that would be interesting, when schedules permit...

Hugo DeGaris:
Will There be Cyborgs?

Interview first published April 2011

My good friend Hugo DeGaris first became prominent for founding the field of "evolvable hardware" – the use of evolutionary algorithms to create new circuitry, using hardware that reconfigures itself dynamically based on software instructions. He developed novel supercomputer technology for evolvable hardware – first in Japan in the late 1980s and early 1990s, and then at a fantastic research company called StarLab in Brussels, which unfortunately crashed in late 2001 during the dot-com boom, not long after the same economic issues killed my first AI company, Webmind Inc.

After a stint at Utah State University, working mostly on quantum computing, Hugo relocated to China, where I visited him frequently -- first at Wuhan University and then at Xiamen University, where I became an adjunct professor and helped him supervise the Artificial Brain Lab. He wrote a rather controversial "science faction" book called *The Artilect War*, foreseeing a huge World War III between Terrans (fearing AI and valuing traditional humanity) and Cosmists (wanting the growth of intelligence throughout the cosmos, even if it means the end of the human race). A little later he began to wonder if there might emerge a third camp as well – the Cyborgists, wanting the best of both worlds... A desire that Hugo considers a wishful delusion.

In 2008-2010 in Xiamen we had some good times working on AGI together, as well as discussing a host of topics, ranging from future science and technology to philosophy, music, literature, politics, you name it. In 2009 we did an AGI Summer School in Xiamen – the first event of its kind, and a heck of a cross-cultural learning experience for all concerned. In mid-2010 Hugo retired from AGI work to focus on studying fundamental physics and associated math, thinking about femtotech and writing futurist and political essays – but since his retirement, I've kept on working with the Xiamen AI lab, though the lab's focus has drifted considerably from Hugo's interests in brain simulation and scalable evolutionary computing. And since he's retired to the outskirts of the city of Xiamen, we get the chance to chat face to face every time I visit Xiamen for research collaboration.

But though Hugo and I enjoy each other's ideas, we definitely don't always see precisely eye to eye. The dialogue comprising this chapter focuses on one of our many disagreements: Hugo's notion that the future will be defined and dominated by artilects – vastly superhuman artificial intellects – to an extent that cyborgs are a downright implausibility!

I do think that incredibly advanced AGI systems are likely to exist, leaving humans and partly-human cyborgs far behind in the virtual dust. But I don't see this as contradictory to the persistence of cyborgs, alongside. I don't really "get" Hugo's confidence that "there will be no cyborgs" – it seems to emanate from some deep intuition that I don't share.

Ben

Hugo, what exactly do you mean when you say "there will be no cyborgs"?

Hugo

My basic point is really simple: "There will be no cyborgs" in the sense that the computational capacity of a nanoteched grain of sand is a <u>quintillion</u> times that of the human brain, making the distinction between a cyborg and a pure artilect pointless. Any human being wishing to convert himself into a cyborg will effectively be killing himself, unless he dramatically impairs and restricts the capability of his cyborg portion. He will no longer be "he" but an "artilect in human disguise." This argument has implications for the species dominance debate.

Ben

OK – so why do you say cyborgization equals death?

Hugo

In my media appearances I often talk about a grain of nanoteched sand being added to a newly born baby's brain to convert it into a cyborg. I finally got around to actually calculating just how much greater the computational capacity of that grain of sand would be compared to that of a human brain (assuming classical computing capacities – if one assumes quantum computing capacities, the quantum estimate is astronomically larger!)

Let's start with some basic assumptions. Let the grain of sand be a 1 mm cube (i.e. 10^{-3} m on a side). Assume the molecules in the sand have a cubic dimension of 1 nm on a side (i.e. 10^{-9} m). Let each molecule consist of 10 atoms (for the purposes of an "order of magnitude" calculation). Assume the grain of sand has been nanoteched such that each atom can switch its state usefully in a femto-second (i.e. 10^{-15} of a second). Assume the computational capacity of the human brain is 10^{16} bits per second (i.e. 100 billion neurons in the human brain, times 10,000, the average number of connections between neurons, times 10, the maximum number of bits per second firing rate at each interneuronal (synaptic) connection = $10^{11}*10^{4}*10^{1} = 10^{16}$.

I will now show that the nanoteched grain of sand has a total bit switching (computational) rate that is a factor of a *quintillion* (a million trillion) times larger than the brain's 10^{16} bits per second. How many sand molecules in the cubic mm? Answer : A million cubed, i.e. 10^{18}, with each of the 10 atoms per molecule switching 10^{15} times per second, so a total switching (bits per second) rate of 10^{18} times 10^{15} times $10^{1} = 10^{34}$. This is $10^{34}/10^{16}$ = 10^{18} times greater, i.e. a million trillion, or a *quintillion*.

Ben

Yes, that's certainly a compelling point!

Now that I think about it, another interesting calculation is obtained by applying the Bekenstein bound[18] from fundamental physics.

You're probably familiar with this, right? Related to black hole thermodynamics, this is a bound on the number of bits of information that can possibly be contained in a certain amount of matter, assuming current physics is correct.

According to the Bekenstein bound the number of bits possibly storable in the matter comprising a human brain is around 10^{42}. Factoring in the smaller diameter and mass of a grain of sand, one decreases this number by a few powers of ten, arriving at an estimate around 10^{35} or so for the sand grain. Compare this to estimates in the range $10^{13} - 10^{20}$ for the human brain, based on our current understanding[19] of psychology and neuroscience Of course, a human brain cannot approach the Bekenstein bound without being restructured so as to constitute some very non-human-brain-like strange matter. A cyborg combining a human brain with a grain of "sand" composed of strange matter that approaches the Bekenstein bound, would potentially contain 10^{35} bits in the femtotech sand grain component, and 10^{21} bits or so bits in the legacy-human-brain component.

Hugo

Right. And this has huge political consequences for the species dominance debate

People like Kurzweil, Warwick etc. claim that there is a third ideological human group in the species dominance debate, besides the Cosmists (pro artilect building), and the Terrans (anti artilect building), namely the Cyborgists, who want to add

---

[18] http://en.wikipedia.org/wiki/Bekenstein_bound
[19] http://www.merkle.com/humanMemory.html

artilectual components to themselves, thus converting themselves into artilect gods. They argue that by converting humans into artilects via the cyborg route, it will be possible to avoid a gigadeath "artilect war" between the Cosmists and the Terrans, for the simple reason that the human race will have converted itself into artilects, so there won't be any Cosmists or Terrans left (i.e. human beings) to wage the war. Everyone will have been upgraded into an artilect.

There's not much point is using the term cyborg, because as the above calculation shows, the human component in the cyborgian brain is only one part in a quintillion (a million trillion), so can effectively be totally ignored. So really, there is no third ideological group. There are only Cosmists, Terrans (and artilects.) Even in the very early days of cyborgian conversion, the vastly superior computational capacities of nanoteched matter will make itself felt. The first cyborgs will very quickly become artilects. Since their bodies would still be human, the Terran paranoia against them would be great. The Terrans would be unable to distinguish a cyborg from a human just by looking at him, hence all human looking beings would be suspects.

A potential cyborg then needs to consider the fact of the above calculation and become fully conscious that a decision to "add the grain of sand (or more)" is a decision to commit suicide as a human being. The resulting cyborg is utterly dominated by the artilectual capacity of the sand grain which totally drowns out the human component (one part in a quintillion, and probably vastly more as quantum computing scales up).

This means that if a mother "cyborgs" her baby, she loses her baby. If adult children cyborg themselves, their elderly parents lose their adult children. Thus it is clear that cyborging will be *profoundly disruptive* to humanity. A large proportion (probably about half, and this needs to be researched more thoroughly, e.g. the BBC vote of 2006 and the Coast to Coast radio poll of 2005) of humanity is deeply, viscerally opposed to Cyborgism and Cosmism, and will go to war to stop both the Cosmists from building pure artilects and the Cyborgists from converting themselves into (near) pure artilects. They will simply *not tolerate* becoming an inferior species, with all the enormous risks to the survival of the human species that the rise of the artilect raises.

When I hear Cyborgists saying such things as "I want to become an artilect god myself, by converting myself into an artilect by adding artilectual components to my brain", I become cynical, because I wonder if they realize how enormous the computational capacity of artilected matter is. I sense that too often, these Cyborgists are ignorant of the "physcomp" (physics of computation) calculations applicable to nanoteched matter. They need to be made aware that a decision to cyborg themselves is a major one – it means the end of their lives as humans. They would no longer be human, they would be artilects, which from a human being's point of view, is to say that they would be utterly, utterly alien, frightening, and to the Terrans, worthy of extermination.

I feel that 2011 will be the year that the "species dominance debate" goes mainstream (at least in the US, where this debate has been led.) Ray Kurzweil has recently been featured in a Times magazine article and I was featured in a History Channel program "Prophets of Doom". There will be several movies in 2011 on the same theme, so the general public will become more exposed to the ideas. If one thinks back 5 years and asks oneself, "How conscious was I of the climate change issue?" Most people would say, "Only a little, if at all." I would not be at all surprised that 5 years from now (2015), the issue of species dominance will be part of the educated person's general knowledge, as much as is climate change today.

And after the issue heats up, the species dominance issue will not go away! Every month technical progress is taking humanity closer to being able to build artilects. The issue will heat up, and tempers will flare. Political parties will be formed on the issue. The debate will later start raging, then the assassinations and sabotages will start. People will take sides. The two ideologies (Cosmism and Terranism) are both very powerful (i.e. building gods vs. preserving the human species — "do we build gods or our potential exterminators?") and so far have split humanity down the middle. Most individuals are ambivalent about the issue — feeling the awe of building artilect gods as well as feeling the horror of a prospective gigadeath artilect war.

Personally, I'm glad I'm alive now. I don't want to see all this horror that is going to consume my grandchild and billions like him. I just don't want to see it. If I live into my 90s, I will see the species dominance debate rage (it's starting already amongst the AI specialists) but I won't see the war. Thank god.

Ben

Hmmm, personally I hope to live long enough to see the Singularity — "live long enough to live forever", as they say! I'm 44 now so I think I might possibly make it. If not, I'm signed up for cryopreservation via Alcor, so my human or robot descendants can defrost me!

About the "no cyborgs" thing — there's certainly a lot of truth to what you say, but it seems to me you're overlooking one major point. Some people may want to become cyborgs with a more *limited* machine component than what is potentially possible, i.e. they might *choose* to be 50% human and 50% machine (in terms of computing power), even though it would technically be possible to become 99.99999% machine (computationally), and only .00001% human. This doesn't affect your arguments about species dominance, but it does IMO mean there could really be three political groups — a) those wanting only legacy humanity (Terrans), b) those wanting only limited expansion of human capabilities, e.g. 50-50 human/machine hybrids (the Cyborgists), c) those wanting expansion without unnecessary limits (the Cosmists). From a Cosmist viewpoint, the Cyborgists and Terrans will be basically the same thing. From a Terran viewpoint, the Cyborgists and Cosmists will be basically the same thing. But to a Cyborgist, the Terrans and Cosmists will seem qualitatively different.

Or is your point mainly that, once we have the femtotech grains of sand available, they're just going to achieve massive superintelligence so quickly that all the distinctions between inferior-intelligence human-scale beings are going to become totally irrelevant?

But don't you think it's possible that the superhuman artilects may let pesky little humans and cyborgs keep on existing and minding their own business, just as we humans do with ants and bacteria?

Hugo

If the "artilectual factor" is only million times more, or a billion, or whatever, the qualitative conclusion is the same, i.e. the human component has been swamped. So there's little point from the Terran viewpoint in distinguishing between an artilect and a cyborg. The so-called "third option" of Cyborgism is a phony one. There really are only 2 options – Terran or Cosmist.

Ben

Hmmm, I just don't quite see it that way. It seems you could have cyborgs with 50% human and 50% machine thought power – if you want to build them. Those would just be very weak cyborgs compared to more powerful artilects. Just like we could now build vehicles propelled by a combination of human and machine power, even though they would be far slower than vehicles propelled by pure machine power.

They would present some dilemmas on their own – for instance, if weak cyborgs were twice as smart and strong as humans, that would certainly cause major social issues. You could imagine a government that made weak cyborgs legal and outlawed strong ones, right?

I agree though that UNLESS you had some kind of all-seeing dictatorial government preventing the development of powerful artilects, this kind of "weak cyborg" would be irrelevant to the species dominance debate. Because artilects will just be a far more significant issue.

Hugo

Right. The last part of your previous question is critical in this whole species dominance debate, namely that of "risk", i.e. the risk that if the artilects come into being, that they may exterminate humans, for whatever reason. I just don't see Globan (my name for the coming world state) Terran politicians around mid-century, tolerating the Cosmists demands to build artilects. To do so would be to accept that the fate of humanity lies with the artilects. Humans would become the "No. 2" species. We would lose control of our own fate. I see the Terran politicians drawing a line in the sand, i.e. pushing hard for a globally legislated maximum AIQ (artificial intelligence quotient) and anyone superseding it is to become a criminal and be prosecuted.

Ben

Right. A maximum AIQ could allow weak cyborgs but not strong ones.

Hugo

The attempt to impose limits like this will force the Cosmists to organize and oppose such a ban. The rhetoric either way will heat up (the "species dominance debate"), and spill over into the "Artilect War", which with 21st century weapons, will lead to "gigadeath".

Ben

That's one potential outcome. It seems there are other possibilities also.

For instance, greater and greater AIQ could gradually sneak up on people. Maybe nobody will want to say no to a 10% increase in cyborg and artilect intelligence. But then those 10% increments keep piling up year after year, till eventually you have massively superhuman intelligences, and nobody raised a big fight about it. Many social changes sneak up gradually in that sort of way.

Or, on the other hand, some Cosmists could create a very powerful artilect and instruct it to prevent any kind of war from happening, with its superhuman capabilities. After that, all bets are off, because who knows what the artilect will do. Unless some future science gives us a way to somehow statistically predict the behavior of our artilectual creations, in spite of their greater intelligence compared to us. And I know you consider this kind of prediction basically a lost cause!

DeGaris Interviews Goertzel: Seeking the Sputnik of AGI

Interview first published March 2011

A couple months after I interviewed Hugo DeGaris for H+ Magazine, Hugo suggested it would be interesting to play a role-reversal game and ask me some interview questions – about my AGI research and my views on the future of humanity and intelligence. It was a typical Ben and Hugo conversation, veering from current technology to future speculation to international politics.

Hugo

About 5 years ago, I was staying at a mutual friend's apartment in Washington DC, just before moving full time to China. At the time you took the view that it would NOT be necessary to have a full knowledge of the human brain to be able to create a human level artificial intelligence. You thought it could be done years earlier using a more humanly engineered approach rather than a "reverse engineering the brain" approach. What are your thoughts on that attitude now, 5 years down the road?

Ben

Wow, was that really 5 years ago? Egads, time flies!! But my view remains the same...

Neuroscience has advanced impressively since then, but more in terms of its understanding of the details than in its holistic vision of the brain. We still don't know exactly how neurons work, we still don't know how concepts are represented in the brain nor how reasoning works, etc. We still can't image the brain with simultaneously high spatial and temporal precision. Etc.

Artificial General Intelligence hasn't advanced as visibly as neuroscience since then, but I think it has advanced. The pursuit of AGI now exists as a well-defined field of research, which wasn't the case back then. And many advances have been made in specific areas of importance for AGI – deep learning models of perception, probabilistic logical inference, automated program learning, scalable graph knowledge stores, and so forth. We also have a vibrant open-source AGI project, OpenCog, which I hope will take off in the next few years the same way Linux did a while back.

Both approaches have a significant ways to go before yielding human-level AGI, but I'd say both have the same basic strengths and weaknesses they did 5 years ago, having advanced steadily but not dramatically.

Hugo

So, which approach do you feel will build human level AI first, your symbolic engineered approach, or reverse engineering of the brain? Why?

Ben

I wouldn't characterize my approach as "symbolic", I think that's a bit of a loaded and misleading term given the history of AI. My approach involves a system that learns from experience. It does include some probabilistic logic rules that are fairly described as "symbolic", but it also includes dynamics very similar to attractor neural nets, and we're now integrating a deep learning hierarchical perception system, etc. It's an integrative experiential learning based approach, not a typical symbolic approach.

Anyway, quibbles over terminology aside, do I think an integrative computer science approach or a brain simulation approach will get there faster?

I think that an integrative computer science approach will get there faster UNLESS this approach is starved of funding and attention, while the brain simulation approach gets a lot of money and effort.

I think we basically know how to get there via the integrative comp sci approach NOW, whereas to follow the neuroscience approach, we'd first need to understand an awful lot more about the brain than we can do with current brain measurement technology. But still, even if one of the current AGI projects – like the OpenCog project I cofounded – is truly workable, it will take dozens of man-years of effort to get to human-level AGI by one of these routes. That's not much in the historical time-scale, but it's a nontrivial amount of human effort to pull together without serious backing from government or corporate sources. Right now OpenCog is funded by a ragtag variety of different approaches, supplemented by the wonderful efforts of some unpaid volunteers – but if this situation continues (for OpenCog and other integrative CS based AGI projects), progress won't be all that fast, and it's not clear which approach will get there first.

What I'm hoping is that, once OpenCog or some other project makes a sufficiently impressive AGI demonstration, there will be a kind of "Sputnik moment" for AGI, and the world will suddenly wake up and see that powerful AGI is a real possibility. And then the excitement and the funding will pour in, and we'll see a massive acceleration of progress. If this AGI Sputnik moment happened in 2012 or 2013 or 2014, for example, then the integrative CS approach would leave the brain simulation approach in the dust – because by that time, we almost surely still won't be able to measure the brain with simultaneously high spatial and temporal precision, so we still won't be able to form an accurate and detailed understanding of how human thinking works.

Hugo

As machines become increasingly intelligent, how do you see human politics unfolding? What are your most probable scenarios? Which do you feel is *the* most probable?

Ben

I see human organizations like corporations and governments becoming gradually more and more dependent on machine intelligence, so that they no longer remember how they existed without it.

I see AI and allied technologies as leading to a lot of awfully wonderful things.

A gradual decrease in scarcity, meaning an end to poverty.

The curing of diseases, including the diseases comprising aging, leading ultimately to radical life extension.

Increased globalization, and eventually a world state in some form (maybe something vaguely like the European Union extended over the whole planet, and then beyond the planet).

The emergence of a sort of "global brain", a distributed emergent intelligence fusing AIs and people and the Net into a new form of mind never before seen on Earth.

Increased openness and transparency, which will make government and business run a lot more smoothly. And will also trigger big changes in individual and collective human psychology. David Brin's writings on sousveillance are quite relevant here, by the way, e.g. the Transparent Society. Also you can look at Wikileaks and the current Mideast revolutions as related to this.

But exactly how all this will play out is hard to say right now, because so much depends on the relative timings of various events. There will be advances in "artificial experts", AI systems that lack humanlike autonomy and human-level general intelligence, but still help solve very important and difficult problems. And then there will be advances in true, autonomous, self-understanding AGI. Depending on which of these advances faster, we'll see different sorts of future scenarios unfold.

If we get super-powerful AGI first, then if all goes well the AGI will be able to solve a lot of social problems in one fell swoop. If we get a lot of artificial experts first, then we'll see problems gradually get solved and society gradually reorganized, and then finally a true AGI will come into this reorganized society.

Hugo

In a recent email to me you said "I don't think it's productive to cast the issue as *species dominance*". Why do you feel that?

Ben

A species dominance war – a battle between humans and AI machines – is one way that the mid-term future could pan out, but we have no reason to think it's the most likely way. And it's possible that focusing on this sort of outcome too much (as many of our science fiction movies have, just because it makes good theater) may even increase the odds of it happening. Sometimes life follows fiction, because the movies someone sees and the books they read help shape their mind.

I find Ray Kurzweil a bit overoptimistic in his view on the future, but maybe his overoptimism is performing a valuable service: By placing the optimistic vision of a "kinder, gentler Singularity" in peoples' minds, maybe he'll help that kind of future to come about. I'd imagine he has thought about it this way, alongside other perspectives….

Another possibility, for example, is that humans may gradually fuse with machines, and let the machine component gradually get more and more intelligent, so that first we have cyborgs with a fairly equal mix of human and machine, and then gradually the machine takes over and becomes the dominant portion. In this case we could feel ourselves become superhuman god-minds, rather than having a (losing) war with superhuman god-minds that are external to ourselves. There would be no species dominance debate, but rather a continuous transition from one "species" into another. And quite possibly the superhuman cyborgs and god-mind AIs would allow legacy humans to continue to exist alongside themselves, just as we allow ants to keep crawling around in the national park, and bacteria to course around inside those ants.

Of course, you could point out that some human beings and some political organizations would be made very mad by the preceding few paragraphs, and would argue to wipe out all the nasty risky techno-geeks who entertain crazy ideas like gradually becoming superhuman god-mind cyborg AIs. So, could there be conflicts between people who like this sort of wild ambitious futurist vision, and those who think it's too dangerous to play with? Of course there could. But focusing on the potential consequences of such conflict seems pointless to me, because they're so unknown at this point, and there are so many other possibilities as well. Maybe this sort of conflict of opinion will someday, somewhere, unfold into a violent conflict or maybe it won't. Maybe Ray Kurzweil is right that the advocates of gradual cyborgization will have vastly more advanced capabilities of defense, offense and organization than their opponents, so that the practical possibility of a really violent conflict between the Cosmists and the Terrans (to use your terminology) won't be there.

After all, right now there is a conflict between people who want to roll back to medieval technology and attitudes (Al Qaeda) and modern technological society – and who's winning? They knocked down the World Trade Center, probably aided in many ways by their connections with the Saudis, who are wealthy because of selling oil to technological nations, and are shielded somewhat by their close connections with the US power elite (e.g. the Bush family). But they're coming nowhere close to winning their war on technological progress and cultural modernization. Our weapons are better – and our memes are stickier. When their kids find out about modern culture and technology, a lot of them are co-opted to our side. When our kids find out about the more violent and anti-technology strains of fundamentalist Islam, relatively few are tempted. My guess is this sort of pattern will continue.

Hugo

Are you mystified by the nature of consciousness?

Ben

Not at all. Consciousness is the basic ground of the universe. It's everywhere and everywhen (and beyond time and space, in fact). It manifests differently in different sorts of systems, so human consciousness is different from rock consciousness or dog consciousness, and AI consciousness will be yet different. A human-like AI will have consciousness somewhat similar to that of a human being, whereas a radically superhumanly intelligent AI will surely have a very different sort of conscious experience.

To me, experience comes first, science and engineering second. How do I know about atoms, molecules, AI and computers, and Hugo DeGaris, and the English language? I know because these are certain patterns of arrangement of my experience, because these are certain patterns that have arisen as explanations of some of my observations, and so forth. The experiential observations and feelings come first, and then the idea and model of the physical world comes after that, built out of observations and feelings. So the idea that there's this objective world out there independent of experience, and we need to be puzzled about how experience fits into it, seems rather absurd to me. Experience is where it all starts out, and everything else is just patterns of arrangement of experience (these patterns of course being part of experience too)...

You could call this Buddhistic or panpsychistic or whatever, but to me it's just the most basic sort of common sense.

So, while I recognize their entertainment value, and their possible value in terms of providing the mind's muscles a cognitive workout — I basically see all the academic and philosophical arguments about consciousness as irrelevancies. The fact that consciousness is a conundrum within some common construals of the modern scientific world view, tells us very little about consciousness, and a lot about the inadequacies of this world view…

Hugo

Do you think humanity will be able to create conscious machines?

Ben

Absolutely, yes.

Hugo

If someone holds a gun to your head and forces you to choose between a god like artilect coming into existence but humanity gets destroyed as a result, OR the artilect is never created, and hence humanity survives, which would you choose and why? Remember the gun at your head.

Ben

Well, I guess none of us knows what we'd really do in that sort of situation until we're in it. Like in the book "Sophie's Choice." But my gut reaction is: I'd choose humanity. As I type these words, the youngest of my three kids, my 13 year old daughter Scheherazade, is sitting a few feet away from me doing her geometry homework and listening to Scriabin Op. Fantasy 28 on her new MacBook Air that my parents got her for Hanukah. I'm not going to will her death to create a superhuman artilect. Gut feeling: I'd probably sacrifice myself to create a superhuman artilect, but not my kids… I do have huge ambitions and interests going way beyond the human race – but I'm still a human.

How about you? What do you reckon you'd choose?

Hugo

I vacillate. When I look at the happy people in the park, I feel Terran. When I stare at astronomy books where each little dot is a galaxy in the famous Hubble "Deep Field" photo, I feel Cosmist. But if I REALLY had to choose, I think I would choose Cosmist. I think it would be a cosmic tragedy to freeze evolution at our puny human level. This is the biggest and toughest decision humanity will ever have to make. "Do we build gods, or do we build our potential exterminators?"

Ben

Well, let's hope we don't have to make that choice. I see no reason why it's impossible to create vastly superhuman minds – and even merge with them – while still leaving a corner of the cosmos for legacy humans to continue to exist in all their flawed ape-like beauty!...

Hugo

How do you see humanity's next 100 years?

Ben

I guess I largely answered this already, right? I see the creation of superhuman AGI during this century as highly likely. Following that, I see a massive and probably irreducible uncertainty. But I think there's a reasonably high chance that what will happen is:

... Some superhuman AGIs, seeded by our creations, will leave our boring little corner of the universe

... Some humans will gradually cyborgify themselves into superhuman AGI god-minds, and probably bid this corner of the Cosmos adieu as well

... Some humans will opt to stay legacy humans, and others will opt to be cyborgs of various forms, with various combinations of human and engineered traits

... The legacy humans and "weak cyborgs" will find their activities regulated by some sort of mildly superhuman "Nanny AI" that prevents too much havoc or destruction from happening

That's my best guess, and I think it would be a pretty nice outcome. But I freely admit I have no strong scientific basis for asserting this is the most probable outcome. There's a hell of a lot of uncertainty about.

Hugo
Do you think friendly AI is possible? Can you justify your answer.

Ben
Do I think it's possible to create AGI systems with vastly superhuman intelligence, that are kind and beneficial to human beings? Absolutely, yes.

Do I think it's possible for humans to create vastly superhuman AGI systems that are somehow provably, guarantee-ably going to be kind and beneficial to human beings? Absolutely not.

It's going to be a matter of biasing the odds.

And the better an AGI theory we have, the more intelligently we'll be able to bias the odds. But I doubt we'll be able to get a good AGI theory via pure armchair theorizing. I think we'll get there via an evolving combination of theory and experiment – experiment meaning, building and interacting with early-stage proto-AGI systems of various sorts.

Hugo
Do you see the US or China being the dominant AI researcher nation in the coming decades?

Ben
Hmmm, I think I'll have to answer that question from two perspectives: a general one, setting aside any considerations related to my own AGI work in particular; and a personal one, in terms of the outlook for my own AGI project.

Generally speaking, my view is that the US has a humongous lead over anywhere else in terms of AGI research. It's the only country with a moderate-sized community of serious researchers who are building serious, practical AGI architectures aimed at the grand goal of human-level intelligence (and beyond). Second place is Europe, not China or India, not even Korea or Japan... The AGI conference series that I co-founded operates every alternate year in the US, and every alternate year elsewhere. The AAAI, the strongest narrow-AI professional organization in the world, is international in scope but US-founded and to a significant extent still US-focused.

The US also has by far the world's best framework for technology transfer – for taking technology out of the lab and into the real world. That's important, because once AGI development reaches a certain point, tech transfer will allow its further development to be funded by the business sector, which has a lot of money. And this kind of thing is hard for other countries to replicate, because it involves a complex ecosystem of interactions between companies of various sizes, universities, and investors of various sorts. It's even hard for cities in the US, outside a certain number of tech hubs, to pull off effectively.

Also, most probably the first powerful AGIs will require a massive server farm, and who's best at doing that? US companies like Google and Amazon and IBM, right? China may have built the world's fastest supercomputer recently, but that's sort of an irrelevancy, because the world doesn't really need supercomputers anymore – what we really need are massive distributed server farms like the ones operated with such stunningly low cost and high efficiency by America's huge Internet companies.

And culturally, the US has more of a culture of innovation and creativity than anywhere else. I know you lived for a while in Utah, which has its pluses but is a very unusual corner of the US – but if you go to any of the major cities or tech hubs, or even a lot of out-of-the-way college towns, you'll see a spirit of enthusiastic new-idea-generation among young adults that is just unmatched anywhere else on the planet. Also a spirit of teamwork, that leads a group of friends just out of college to start a software company together, cooperating informally outside the scope of any institution or bureaucracy.

Look at any list of the most exciting tech companies or the biggest scientific breakthroughs in the last few years, and while it will look plenty international, you'll see a lot of US there. Many of the US scientists and technologists will have non-Anglo-Saxon-sounding names – including many that are Chinese or Indian – but that's part of the US's power. Many of the best students and scientists from around the world come to America to study, or teach, or do research, or start companies, etc. That's how the US rose to science and engineering prominence in the first place – not through descendants of the Puritans, but through much more recent immigrants. My great-grandparents were Eastern European Jews who immigrated to the US in the first couple decades of the last century. They were farmers and shopkeepers in Europe, now their descendants are scientists and professors, executives and teachers, etc. This is same sort of story that's now bringing so many brilliant Asians to America to push science and technology forward.

So, hey, God bless America! What more can I say....?

I've lived a lot of places in my life (Brazil, Oregon, New Jersey, Philly, four of the five boroughs of New York City, Australia, New Zealand, New Mexico) – but, among a lot of more exotic locales, I spent 9 years of my career (from 2003 thru early 2011) living near Washington DC. [20] During that interval I was being a bit of a

---

[20] This interview was done in 2011. In late 2011 Ben Goertzel relocated to Hong Kong, where he's lived since that time, doing various sorts of AI R&D, funded with a combination of private financing and Hong Kong government funding. Also, relatedly, in 2013 Ben Goertzel helped Getnet Aseffa, an Ethiopian roboticist and entrepreneur, to launch iCog Labs, the first AI development firm in Ethiopia. The international landscape of AI and AGI research continues to shift in interestin ways.

"Beltway bandit," I suppose. I lived a few miles from the National Institutes of Health, for which I did a lot of bioinformatics work; and I also did some AI consulting for various companies working with other government agencies. DC has its pluses and minuses. I wouldn't say I fit into the culture of DC too naturally; but there's a lot more interesting R&D going on in DC than most people realize, because the culture isn't publicity-oriented. And in some ways there's a longer-term focus in DC than one finds in Silicon Valley, where there's so much obsession with moving super-fast and getting profits or cash flow or eyeballs or whatever as quickly as possible... The Chinese government thinks 30 years ahead (one of its major advantages compared to the US, I might add), Wall Street thinks a quarter ahead, Silicon Valley thinks maybe 3 years ahead (Bay area VCs typically only want to invest in startups that have some kind of exit strategy within 3 years or so; and they usually push you pretty hard to launch your product within 6 months of funding – a default mode of operation which is an awkward fit for a project like building AGI), and DC is somewhere between Silicon Valley and China...

But still ... having said all that ... There's always another side to the coin, right? On the other hand, if -- if, if, if – the US manages to squander these huge advantages during the next few decades, via pushing all its funding and focus on other stuff besides AGI and closely allied technologies ... Then who knows what will happen. Economically, China and India are gradually catching up to the US and Europe and Korea and Japan... They're gradually urbanizing and educating and modernizing their huge rural populations. And eventually China will probably adopt some sort of Western style democracy, with free press and all that good stuff, and that will probably help Chinese culture to move further in the direction of free expression and informal team work and encouragement of individual creativity – things that I think are extremely important for fostering progress in frontier areas like AGI. And eventually India will overcome its patterns of corruption and confusion and become a First World country as well. And when these advances happen in Asia, then maybe we'll see a more balanced pattern of emigration, where as many smart students move from the US to Asia as vice versa. If the advent of AGI is delayed till that point – we're talking maybe 2040 or so I would reckon – then maybe China or India is where the great breakthrough will happen.

I do think China is probably going to advance beyond the US in several areas in the next couple decades. They're far, far better at cheaply making massive infrastructure improvements than we are. And they're putting way more effort and brilliance into energy innovations than we are. To name just two examples. And then there's stem cell research, where the US still has more sophistication, but China has fewer regulatory slowdowns; and other areas of biomedical research where they excel. But these areas are largely to do with building big stuff or doing a lot of experimentation. I think the Chinese can move ahead in this sort of area more easily than in something like AGI research. I think AGI research depends mostly on the closely coordinated activity of small informal or semi-formal groups of people pursuing oddball ideas, and I don't think this is what Chinese culture and institutions are currently best at fostering.

Another factor acting against the USA, is that the US AI research community (along with its research funding agencies) is largely mired in some unproductive ideas, the result of the long legacy of US AI research. And it's true that the Chinese research community and research funders aren't similarly conceptually constricted – they have fewer unproductive conceptual biases than US AI researchers, on the whole. But if you look at the details, what most Chinese academics seem to care most about these days is publishing papers in SCI-indexed journals and getting their citation counts higher – and the way to do this is definitely NOT to pursue long-term oddball speculative AGI research....

You might be able to frame an interesting argument in favor of India as a future AGI research center, on this basis. They seem a bit less obsessed with citation counts than the Chinese, and they have a long history of creative thinking about mind and consciousness, even longer than the Chinese! Modern consciousness studies could learn a lot from some of the medieval Indian Buddhist logicians. Plus a lot of Silicon Valley's hi-tech expertise is getting outsourced to Bangalore. And the IITs are more analogous to top-flight US technical universities than anything in China – though Chinese universities also have their strengths. But anyway, this is just wild speculation, right? For now there's no doubt that the practical nexus of AGI research remains in America (in spite of lots of great work being done in Germany and other places). AGI leadership is America's to lose... And it may well lose it, time will tell... Or America-based AGI research may advance sufficiently fast that nobody else has time to catch up...

Hugo

OK, that was your general answer... Now what about your personal answer? I know you've been spending a lot of time in China lately, and you're working with students at Xiamen University in the lab I ran there before I retired, as well as with a team in Hong Kong....

Ben

Yeah, that was my general answer. Now I'll give my personal answer – that is, my answer based on my faith in my own AGI project.

I think that the OpenCog project, which I co-founded, is on an R&D path that has a fairly high probability of leading to human-level general intelligence (and then beyond). The basic ideas are already laid out in some fairly careful (and voluminous) writing, and we have a codebase that already functions and implements some core parts of the design, and a great team of brilliant AGI enthusiasts who understand the vision and the details... So, if my faith in OpenCog is correct, then the "US versus China" question becomes partly a question of whether OpenCog gets developed in the US or China.

Interestingly, it seems the answer is probably going to be: both!... And other places too. It's an open source project with contributors from all over the place.

My company Novamente LLC is driving part (though by no means all) of OpenCog development, and we have some programmers in the US contributing to OpenCog based on US government contracts (which are for narrow-AI projects that use OpenCog, rather than for AGI per se), as well as a key AGI researcher in Bulgaria, and some great AI programmers in Belo Horizonte, Brazil, whom I've been working with since 1998. There's also a project at Hong Kong Polytechnic University, co-sponsored by the Hong Kong government's Innovation in Technology Fund and Novamente LLC, which is applying OpenCog to create intelligent game characters. And there's a handful of students at Xiamen University in China working on making a computer vision front end for OpenCog, based on Itamar Arel's DeSTIN system (note that Itamar is from Israel, but currently working in the US, as a prof at the University of Tennessee Knoxville, as well as CTO of a Silicon Valley software company, Binatix). Now, the AI programmers on the Hong Kong project consist of two guys from New Zealand (including Dr. Joel Pitt, the technical lead on the project) and also three exchange students from Xiamen University. In April I'll be spending a few weeks in Hong Kong with the team there, along with Dr. Joscha Bach from Germany.

My point in recounting all those boring details about people and places is – maybe your question is just too 20th century. Maybe AGI won't be developed in any particular place, but rather on the interwebs, making use of the strengths of the US as well as the strengths of China, Europe, Brazil, New Zealand and so on and so forth.

Or maybe the US or Chinese government will decide OpenCog is the golden path to AGI and throw massive funding at us, and we'll end up relocating the team in one location – it's certainly possible. We're open to all offers that will allow us to keep our code open source!

So far I have found the Chinese research funding establishment, and the Chinese university system, to be much more open to radical new approaches to AGI research than their American analogues. In part this is just because they have a lot less experience with AI in general (whether narrow AI or AGI). They don't have any preconceived notions about what might work, and they don't have such an elaborate "AI brain trust" of respected older professors at famous universities with strong opinions about which AI approaches are worthwhile and which are not. I've gotten to know the leaders of the Chinese AI research community, and they're much much more receptive to radical AGI thinking than their American analogues. Zhongzhi Shi from the Chinese Academy of Sciences is going to come speak about Chinese AGI efforts at the AGI-11 conference in California in August — and I've also had some great conversations with your friend Yixin Zhang, who's the head of the Chinese AI Association. I went to their conference last year in Beijing, and as you'll recall our joint paper on our work with intelligent robots in Xiamen won the Best Paper prize! At the moment their efforts are reasonably well funded, but not to the level of Chinese work on semiconductors or supercomputers or wind power or stem cell research, etc. etc. But certainly I can see a possible future where some higher-ups in the Chinese government decide to put a massive amount of money into intelligent robotics or some other AGI application, enough to tempt a critical mass of Western AGI researchers as well as attract a lot of top Chinese students... If that does happen, we could well see the world's "AGI Sputnik" occur in China. And if this happens, it will be interesting to see how the US government responds — will it choose to fund AGI research in a more innovation-friendly way than it's done in the past, or will it respond by more and more aggressively funding the same handful of universities and research paradigms it's been funding since the 1970s?

So overall, putting my general and personal answers together – I feel like in the broad scope, the AGI R&D community is much stronger in the US than anywhere else, and definitely much much stronger than in China. On the other hand, AGI is the sort of thing where one small team with the right idea can make the big breakthrough. So it's entirely possible this big breakthrough could occur outside the US, either via natively-grown ideas, or via some other country like China offering a favorable home to some American-originated AGI project like OpenCog that's too radical in its conceptual foundations to fully win the heart of the US AI research funding establishment.

But ultimately I see the development of AGI in an international context as providing higher odds of a beneficial outcome, than if it's exclusively owned and developed in any one nation. So as well as being an effective way to get work done, I think the international open-source modality we're using for OpenCog is ultimately the most ethically beneficial way to do AGI development...

Well, what do you think? You live in China... I've spent a lot of time there in recent years (and plan to spend a few months there this year), but not as much as you. And you speak the language; I don't. Do you think I'm missing any significant factors in my analysis?

Hugo

I put more emphasis on Chinese economics and national energy. Americans have become fat and complacent, and are not growing economically at anywhere near the Chinese rate. The historical average US economic growth rate is 3%, whereas China's is 10% (and has been sustained pretty much for 30 years). Doing the math, if this incredible energy of the Chinese can be sustained for a few more decades, it will put the rich eastern Chinese cities at a living standard well above that of the US, in which case they can afford to attract the best and most creative human brains in the world to come to China. The US will then see a reverse brain drain, as its best talent moves to "where its at", namely China. With a million talented Westerners in China within a decade, they will bring their "top world" minds with them and shake up China profoundly, modernizing it, legalizing it, democratizing it and civilizing it. Once China finally goes democratic and with its rich salaries, it doesn't matter whether the Chinese can be creative or not. The presence of the best non-Chinese brains in China will ensure an explosion of creativity in that part of the world.

Ben

Hmmm... Chinese economic growth is indeed impressive – but of course, it's easier to grow when you're urbanizing and modernizing a huge rural population. To an extent, the US and Europe and Japan are growing more slowly simply because they've already urbanized and modernized. I guess once China and India have finished modernizing their growth rates may look like those in the rest of the world, right? So your projection that Chinese growth will make Chinese cities richer than US cities may be off-base, because most of Chinese growth has to do with bringing more and more poor people up to the level of the international middle class. But I guess that's a side point, really...

About creativity... Actually, I know many fantastically creative Chinese people (including some working on OpenCog!) and I guess you do too – what seems more lacking in China is a mature ecosystem for turning wacky creative ideas into novel, functional, practical realizations. I'm sure that will come to China eventually, but it requires more than just importing foreigners, it may require some cultural shifts as well – and it's hard to estimate the pace at which those may happen. But China does have the capability to "turn on a dime" when it wants to, so who knows!

About Americans being fat and complacent – hmmm, well I'm a little heavier than I was 20 years ago, but I haven't become a big fat capitalist pig yet... And I don't consider myself all that complacent! Generalizations are dangerous, I guess. San Fran,

Silicon Valley, New York, DC, Boston, Seattle, LA – there's a lot of energy in a lot of US cities; a lot of diversity and a lot of striving and energy. But yeah, I see what you mean – America does sort of take for granted that it's on top, whereas China has more of an edge these days, as if people are pushing extra hard because they know they're coming from behind...

Look at the San Fran Bay area as an example. Sometimes the Silicon Valley tech scene seems a bit tired lately, churning out one cookie-cutter Web 2.0 startup after another. But then the Shanghai startup scene is largely founded on churning out Chinese imitations of Silicon Valley companies. And then you have some really innovative stuff going on in San Fran alongside the Web 2.0 copycat companies, like Halcyon Molecular (aiming at super-cheap DNA sequencing) or Binatix or Vicarious Systems (deep learning based perception processing, aiming toward general intelligence). You don't have a lot of startups in China with that level of "mad science" going on in them, at least not at this point. But maybe you will in 5 or 10 years, maybe in Shanghai or Hong Kong... So there's a lot of complexity in both the US and China, and it's far from clear how it will all pan out. Which is one reason I'm happy OpenCog isn't tied to any one city or country, of course...

Although, actually, now that I mull on it more, I'm starting to think your original question about the US versus China may be a bit misdirected. You seem to have a tendency to see things in polarized, Us versus Them terms, whereas the world may operate more in terms of complex interpenetrating dynamical networks. The dichotomy of Terrans versus Cosmists may not come about because AGIs and nanotech and such may interweave into peoples' bodies and lives so much, step by step, that the man/machine separation comes not to mean much of anything anymore. And, the dichotomy of US versus China may not mean exactly what you think it does. Not only are the two economies bound very tightly together on the explicit level, but there may be more behind-the-scenes political interdependencies than you see in the newspapers. Note that the Chinese government invests masses of money in energy technology projects, in collaboration with Arab investment firms allied with various Saudi princes; and note also that various Saudi princes are closely allied with the Bush family and the whole social network of US oil executives and military/intelligence officials who have played such a big behind-the-scenes role in US politics in the last decade (see Family of Secrets[21] for part of this story, though I'm not saying I fully believe all the author's hypotheses). So maybe the real story of the world today isn't so much about nation versus nation, but more about complex networks of powerful individuals and organizations, operating and collaborating behind the scenes as much as in the open. So then maybe the real question isn't which country will develop AGI, but rather whether it will be developed in service of the oligarchic power-elite network, or in service of humanity at large. And that in itself is a pretty powerful argument for the open-source approach to AGI which, if pursued fully and enthusiastically, allies AGI with a broadly distributed and international network of scientists, engineers and ordinary people. Note that the power elite network doesn't always have to win — it wanted Mubarak to stay in power in Egypt, but that didn't happen, because another, more decentralized and broader, social network proved more powerful.

But we're straying pretty far afield! Maybe we'd better get back to AGI!

Hugo

Yes, politics can certainly be a distraction. But it's an important one, because that's the context in which AGI will be deployed, once it's created.

---

[21] http://www.familyofsecrets.com/

Ben

Indeed! And given the strong possibility for very rapid advancement of AGI[22] once it reaches a certain level of maturity, the context in which it's initially deployed may make a big difference...

Hugo

But, getting back to AGI... Can you list the dominant few ideas in your new book "Engineering General Intelligence[23]".

Ben

Uh oh, a hard question! It was more fun blathering about politics...

That book is sort of a large and unruly beast. It's over 1000 pages and divided into two parts. The first part outlines my general approach to the problem of building advanced AGI, and the second part reviews the OpenCog AGI design – not at the software code level, but at the level of algorithms and knowledge representations and data structures and high level software design.

Part I briefly reviews the "patternist" theory of mind I outlined in a series of books earlier in my career, and summarized in *The Hidden Pattern* in 2006. Basically, a mind is a system of patterns that's organized into a configuration that allows it to effectively recognize patterns in itself and its world. It has certain goals and is particularly oriented to recognize patterns of the form "If I carry out this process, in this context, I'm reasonably likely to achieve this goal or subgoal." The various patterns in the mind are internally organized into certain large-scale networks, like a hierarchical network, and an associative hierarchy, and a reflexive self. The problem of AGI design then comes down to: how do you represent the patterns, and via what patterned

---

[22] http://hplusmagazine.com/2011/03/07/why-an-intelligence-explosion-is-probable/

[23] http://wiki.opencog.org/w/Building_Better_Minds

processes does the pattern system recognize new patterns? This is a pretty high-level philosophical view but it's important to start with the right general perspective or you'll never get anywhere on the AGI problem, no matter how brilliant your technical work nor how big your budget.

Another key conceptual point is that AGI is all about resource limitations. If you don't have limited spacetime resources then you can create a super-powerful AGI using a very short and simple computer program. I pointed this out in my 1993 book The Structure of Intelligence (and others probably saw it much earlier, such as Ray Solomonoff), and Marcus Hutter rigorously proved it in his work on Universal AI a few years ago. So real-world AGI is all about: how do you make a system that displays reasonably general intelligence, biased toward a certain set of goals and environments, and operates within feasible spacetime resources. The AGIs we build don't need to be biased toward the same set of goals and environments that humans are, but there's got to be some overlap or we won't be able to recognize the system as intelligent, given our own biases and limitations.

One concept I spend a fair bit of time on in Part I is cognitive synergy: the idea that a mind, to be intelligent in the human everyday world using feasible computing resources, has got to have multiple somewhat distinct memory stores corresponding to different kinds of knowledge (declarative, procedural, episodic, attentional, intentional (goal-oriented))... And has got to have somewhat different learning processes corresponding to these different memory stores... And then, these learning processes have got to synergize with each other so as to prevent each other from falling into unproductive, general intelligence killing combinatorial explosions.

In the last couple months, my friend and long-time collaborator (since 1993!) Matt Ikle' and I put some effort into formalizing the notion of cognitive synergy using information geometry and related ideas. That went into Engineering General Intelligence too – one of my jobs this month is to integrate that material into the manuscript. We take our cue from general relativity theory, and look at each type of memory in the mind as a kind of curved mindspace, and then look at the combination of memory types as a kind of composite curved mindspace. Then we look at cognition as a matter of trying to follow short paths toward goals in mindspace, and model cognitive synergy as cases where there's a shorter path through the composite mindspace than through any of the memory type specific mindspaces. I'm sort of hoping this geometric view can serve as a unifying theoretical framework for practical work on AGI, something it's lacked so far.

Then at the end of Part I, I talk about the practical roadmap to AGI – which I think should start via making AGI children that learn in virtual-world and robotic preschools. Following that we can integrate these toddler AGIs with our narrow-AI programs that do things like biological data analysis and natural language processing, and build proto-AGI artificial experts with a combination of commonsense intuition and specialized capability. If I have my way, the first artificial expert may be an artificial biologist working on the science of life extension, following up the work I'm doing now with narrow AI in biology with Biomind LLC and Genescient Corp. And then finally, we can move from these artificial experts to real human-level AGIs. This developmental approach gets tied in with ideas from developmental psychology, including Piaget plus more modern ideas. And we also talk about developmental ethics – how you teach an AGI to be ethical, and to carry out ethical judgments using a combination of logical reason and empathic intuition. I've always felt that just as an AGI can ultimately be more intelligent than any human, it can also be more ethical – even according to human standards of ethics. Though I have no doubt that advanced AGIs will also advance beyond humans in their concept of what it is to be ethical.

That's Part I, which is the shorter part. Part II then goes over the OpenCog design and some related technical ideas, explaining a concrete path to achieving the broad concepts sketched in Part I. I explain practical ways of representing each of the kinds of knowledge described in Part I – probabilistic logic relations for declarative knowledge, programs in a simple LISP-like language for procedural knowledge, attractor neural net like activation spreading for attentional knowledge, "movies" runnable in a simulation engine for episodic knowledge, and so forth. And then I explain practical algorithms for dealing with each type of knowledge – probabilistic logical inference and concept formation and some other methods for declarative knowledge; probabilistic evolutionary program learning (MOSES) for procedural knowledge economic attention networks for attentional knowledge; hierarchical deep learning (using Itamar Arel's DeSTIN algorithm) for perception; etc. And I explain how all these different algorithms can work together effectively, helping each other out when they get stuck – and finally, how due to the interoperation of these algorithms in the context of controlling an agent embodied in a world, the mind of the agent will build up the right internal structures, like hierarchical and heterarchical and self networks.

I'm saying "I" here because the book represents my overall vision, but actually I have two co-authors on the book – Nil Geisweiller and Cassio Pennachin – and they're being extremely helpful too. I've been working with Cassio on AI since 1998 and he has awesomely uncommon common sense and a deep understanding of both AI, cog sci and software design issues. And Nil also thoroughly understands the AGI design, and is very helpful at double-checking and improving formal mathematics (I understand math very well, that was my PhD area way back when, but I have an unfortunate tendency to make careless mistakes...). The two of them have written some parts and edited many others; and there are also co-authors for many of the chapters, who have contributed significant thinking. So the book is really a group effort, orchestrated by me but produced together with a lot of the great collaborators I've been luck to have in the last decade or so.

Now, so far our practical work with OpenCog hasn't gotten too far through the grand cosmic theory in the Engineering General

Intelligence book. We've got a basic software framework that handles multiple memory types and learning processes, and we have initial versions of most of the learning processes in place, and the whole thing is built pretty well in C++ in a manner that's designed to be scalable (the code now has some scalability limitations, but it's designed so we can make it extremely scalable by replacing certain specific software objects, without changing the overall system). But we've done only very limited experimentation so far with synergetic interaction between the different cognitive processes. Right now the most activity on the project is happening in Hong Kong, where there's a team working on applying OpenCog to make a smart video game character. We're going to get some interesting cognitive synergy going in that context, during the next couple years

The argument some people have made against this approach is that it's too big, complex and messy. My response is always: OK, and where exactly is your evidence that the brain is not big, complex and messy? The OpenCog design is a hell of a lot simpler and more elegant than the human brain appears to be. I know a fair bit of neuroscience, and I've done some consulting projects where I've gotten to interact with some of the world's greatest neuroscientists – and everything I learn about neuroscience tells me that the brain consists of a lot of neuron types, a lot of neurotransmitter types, a lot of complex networks and cell assemblies spanning different brain regions which have different architectures and dynamics and evolved at different times to meet different constraints. The simplicity and elegance that some people demand in an AGI design, seems utterly absent from the human brain. Of course, it's possible that once we find the true and correct theory of the human brain, the startling simplicity will be apparent. But I doubt it. That's not how biology seems to work.

I think we will ultimately have a simple elegant theory of the overall emergent dynamics of intelligent systems. That's what I'm trying to work toward with the ideas on curved mindspace that I mentioned above. Whether or not those exact ideas are right, I'm sure some theory of that general nature is eventually going to happen. But the particulars of achieving intelligence in complex environments using feasible computational resources – I feel that's always likely to be a bit messy and heterogeneous, involving integration of different kinds of memory stores with different kinds of learning processes associated with them. Just like the theory of evolution is rather simple and elegant, and so is the operation of DNA and RNA -- but the particulars of specific biological systems are always kind of complex and involved.

I'm sure we didn't get every detail right in Engineering General Intelligence – but we're gradually pulling together a bigger and bigger community of really smart, passionate people working on building the OpenCog system, largely inspired by the ideas in that book (plus whatever other related ideas team members bring in, based on their own experience and imagination!). The idea is to be practical and use Engineering General Intelligence and other design ideas to create a real system that does stuff like control video game characters and robots and biology data analysis systems, and then improve the details of the design as we go along. And improve our theories as we go along, based on studying the behaviors of our actual systems. And once we get sufficiently exciting behaviors, trumpet them really loud to the world, and try to create an "AGI Sputnik Moment", after which progress will really accelerate.

And by the way — just to pull the politics and "future of humanity" thread back into things — there's one thing that the theory in Engineering General Intelligence doesn't tell us, which is what goals to give our AGI systems as they develop and learn. If I'm right that Friendly AI is achievable, then crafting this goal system is a pretty important part of the AGI task. And here is another place here some variant of the open source methodology may come in. There's a whole movement toward open governance — the Open Source Party[24] that RU Sirius described in his

---

[24] https://sites.google.com/site/opensourceparty/

recent H+ Magazine article[25], and metagoverment software[26], and so forth. Maybe the goal system for an advanced AGI can be designed by the people of the world in a collaborative way. Maybe some kind of metagoverment software could help us build some kind of "coherent aggregated volition [27]" summarizing the core ideals of humanity, for embodiment in AGI systems. I'd rather have that than see any specific government or company or power-elite network craft the goal system of the first transhuman AGI. This is an area that interests me a lot, though it's very undeveloped as yet....

Anyway there's a lot of work to be done on multiple fronts. But we're getting there, faster than most people think. Getting... Let us hope and strive... To a really funky positive amazing future, rather than an artilect war or worse!

---

[25] http://hplusmagazine.com/2011/02/24/open-source-party-2-0-liberty-democracy-transparency/
[26] http://metagovernment.org/wiki/Main_Page
[27] http://multiverseaccordingtoben.blogspot.gr/2010/03/coherent-aggregated-volition-toward.html

Linas Vepstas: AGI, Open Source and Our Economic Future

Interview first published June 2011

The open source approach to software development has proved extremely productive for a large number of software projects spanning numerous domains. The Linux operating system, which powers 95% of webservers hosting the Internet, is perhaps the best-known example. Since the creation of powerful Artificial General Intelligence is a large task that doesn't fit neatly into the current funding priorities of corporations or government research funding agencies, pursuing AGI via the open-source methodology seems a natural option.

Some have raised ethical concerns about this possibility, feeling that advanced AGI technology (once it's created) will be too dangerous for its innards to be exposed to the general public in the manner that open source entails. On the other hand, open source AGI advocates argue that there is greater safety in the collective eyes and minds of the world than in small elite groups.

I'm hardly a neutral party in this matter, being a co-founder of a substantial open-source AGI project, OpenCog[28], myself. And one of the benefits of involvement with open source software projects that I've found, is the opportunity it provides to interact with a huge variety of interesting people. My intersection with Dr. Linas Vepstas via our mutual engagement with the OpenCog[29] open-source AGI project is a fine example of this. A few years back, Linas and I worked together on a commercial application of OpenCog's natural language toolkit; and since then, as his professional career has moved on, Linas has continued to be a major contributor to OpenCog's codebase and conceptual

---

[28] http://opencog.org/
[29] http://opencog.org/

framework. One dimension of this has been his work maintaining and improving the link parser[30], an open source syntax parsing system originally developed at Carnegie-Mellon University[31], which is wrapped up inside OpenCog's natural language comprehension system. He has also contributed mightily to MOSES, OpenCog's probabilistic evolutionary program learning subsystem, that was originally created by Moshe Looks (now at Google) in our collaborative work that resulted in Moshe's 2006 PhD thesis.

Linas's career started out in physics, with a PhD from SUNY Stony Brook for a thesis on "Chiral Models of the Nucleon," but since that time he's had a diverse career in software development, management and research, including multiple stints at IBM in various areas including graphics and firmware development, and CEO and CTO positions at startups. He was also the lead developer of a major open-source project, the GnuCash[32] free accounting software, and has been very active in Wikipedia[33], creating or majorly revising over 400 mathematics Wikipedia articles, along with dozens of Wikipedia articles in other areas. Currently he's employed at Qualcomm as a Linux kernel developer, and also participating in several other open source projects, including OpenCog. As he puts it, "I'm crazy about open source! Can't live without it!"

Beyond his technical work, Linas's website[34] contains a variety of essays on a range of themes, including sociopolitical ones. One of his persistent foci is the broader implications of open source, free software and related concepts; e.g. in "Is Free Software Inevitable?"[35] he argues that "Free Software is not just a (political) philosophy, or a social or anthropological movement,

---

[30] http://www.abisource.com/projects/link-grammar
[31] http://www.link.cs.cmu.edu/link/
[32] http://www.gnucash.org/
[33] http://en.wikipedia.org/wiki/User:Linas
[34] http://linas.org/%5d
[35] http://linas.org/theory/freetrade.html

but an economic force akin to the removal of trade barriers, we can finally understand why its adoption by corporations promises to be a sea-change in the way that software is used."

Given the deep thinking he's put into the topic, based on his wide experience, it seemed apropos to interview Linas about one of the topics on my mind these days – the benefits and possible costs of doing AGI development using the open source modality.

Ben

Based on your experience with OpenCog and your general understanding, what are the benefits you see of the open-source methodology for an AGI project, in terms of effectively achieving the goal of AGI at the human level and ultimately beyond?

Linas

To properly discuss AGI, one must distinguish between "what AGI is today" from "what we hope AGI might be someday". Today, and certainly in the near term future, AGI is not unlike many other technology projects. This means that the same pros and cons, the same arguments and debates, the same opinions and dynamics apply. Open source projects tend to move slower than commercial projects; they're often less focused, but also less slap-dash, sloppy and rushed. On the other hand, commercially valuable projects are well-funded, and can often be far more sophisticated and powerful than what a rag-tag band of volunteers can pull together.

Open source projects are usually better structured than university projects, for many reasons. One reason is the experience level of the programmers: A grad student may be smart; a professor may be brilliant, but in either case, both are almost surely novice programmers. And novices are, as always, rather mediocre as compared to the seasoned professional. When open source projects are big enough to attract experienced programmers, then the resulting focus on quality, maintainability and testability benefits the project.

All projects, whether closed-source, university, or open source, benefit from strong leadership, and from a team of "alpha coders": When these leave, the project often stalls. When a grad student leaves, whatever source code they wrote goes into almost immediate decline: it bit-rots, it doesn't run on the latest hardware and OS'es, and one day, no longer compiles. Open source projects have the breadth of interest to keep things going after the departure of a leader programmer.

These kinds of points and arguments have been made many times, in many places, during discussions of open source. Many more can be made, and they would generically apply to OpenCog.

Ben

Yeah, there are many similarities between open-source AGI and other OSS projects. But it seems there are also significant differences. In what ways would you say an AGI project like OpenCog differs from a typical OSS project?

Linas

In many ways, OpenCog is quite unlike a typical OSS project. Its goals are diffuse and amorphous. It's never been done before. Most open-source projects begin with the thought "I'm gonna do it just like last time (or just like project X), except this time, I'll make it better, stronger, I'll do it right." AGI has never been done before. The "right way" to do it is highly contested. That makes it fundamentally "research". It also makes it nearly impossible for average software programmers to participate in the project. There's simply too much theory and mathematics and too many obscure bleeding-edge concepts woven into it to allow the typical apps programmer to jump in and help out.

Consider an open-source project focused on music, or accounting, or database programming, or 3D graphics, or web-serving. There are hundreds of thousands of programmers, or more, with domain expertise in these areas. They've "done it before". They can jump into an open-source project, and be productive contributors within hours if not minutes. The number of experienced AGI programmers is miniscule. There's no pool to draw from; one must be self-taught, and the concepts that need to be learned are daunting, and require years of study. Neither OpenCog, nor any other open source AGI project, is going to derive the typical benefits of open participation because of this.

Perhaps OpenCog might be comparable to gcc or the LLVM compilers: the theory of compilation is notoriously complex and arcane, and pretty much requires PhD-level experience in order to participate at the highest levels. On the other hand, the theory of compilers is taught in classes, and there are textbooks on the topic. This cannot be said for AGI. If OpenCog seems to be slow compared to some other open-source projects, these are the reasons why: It is fundamentally difficult.

More interesting, though, perhaps, is to ask what other projects should an open-source AGI project be compared to? Clearly, there are some other overtly AGI-ish projects besides OpenCog out there, and these are more or less directly comparable. But what about things that aren't? A self-driving car is not an AGI project, but certainly solves many of the problems that an AGI project is interested in. The funding, and development staff, for such a project, is certainly much much larger than that for AGI projects. What about automatic language translation? AI avatars in video games? Might such a project "accidentally" evolve into an AGI project, by dint of cash-flow? When one has (tens of) thousands of paid employees, management, and executives working towards some end, it is not uncommon to have skunk-works, research-y, high-risk projects, taking some pie-in-the-sky gamble pursuing some wild ideas. Personally, I believe that there is a very real possibility that one of these areas will sprout some AGI project, if not soon, then certainly in the upcoming years or decade. Generalizing the AI "problem" is just too commercially valuable to not take the risky plunge. It will happen. And, mind you, the "winner take all" meme is a powerful one.

Ben

Hmmm... I see what you're saying but yet I have to say I'm a bit skeptical of a narrow-AI OSS project "turning into AGI". I tend to think that AGI is going to require a large bit of work that is specifically AGI-focused rather than application-focused.

Thinking about OpenCog in particular – I can see that once it gets a bit further along, one could see some extremely interesting hybridizations between OpenCog and some application-focused OSS projects... I can certainly see a lot of promise there! But that would be leveraging OpenCog work that was done specifically with AGI in mind, not wholly under the aegis of pursuing some particular application.

Linas

What I meant to say was corporations with large narrow-AI projects are likely to have small (secret/proprietary) skunk-works AGI projects.

Ben

Yes, that seems plausible. But I suppose these are very rarely open-source projects, and then generally if they fail to bear dramatic fruit quickly, they'll probably end up being dropped or advancing extremely slowly in researchers' spare time. Without naming any names, I know of a couple AGI skunkworks projects in large corporation that fit exactly this description. They're good projects led by great researchers, but they don't get much time or resource allocation from their host companies because they don't show promise of leading to practical results quickly enough. And because they're not open source they can't easily take advantage of the community of AI programmers outside the host company.

Linas

Indeed, my experience with skunk-works projects are that they are very unlikely to be open source, and are unlikely to ever see the light of day as a product or a part of a product; even if an impressive demo or two is created, even if some executive shows interest in it for a while, they tend to dry up and wither. There are occasional exceptions; sometimes the wild idea proves strong enough to become a real-life product.

So all I meant to say is that any corporation with more than a few dozen programmers is going to have some secret lazy Friday afternoon project, and in bigger companies, these can even become full-time funded projects, with 2, 3, 5 or 6 people on them. They go on for a while, and if successful, they morph. I figure the same dynamic happens in narrow-AI companies, and that the engineers at a narrow-AI company are likely to take some random stab at some vaguely AGI-like project. But no, not open source.

Ben

I think AGI is a bigger problem than skunk-works efforts like that will be able to address, though. So if a big company isn't willing to go OSS with its skunk-works AGI project, then it will need to substantially staff and fund it to make real progress. And so far as I can tell that hasn't happened yet. Although, the various AGI skunk-works projects in existence probably still advance the state of the art indirectly in various ways.

Linas

Yes, exactly.

Ben

Now let me shift directions a bit. Some people have expressed worries about the implications of OSS development for AGI ethics in the long term. After all, if the code for the AGI is out there, then it's out there for everyone, including bad guys. On the other hand, in an OSS project there are also generally going to be a lot more people paying attention to the code to spot problems. How do you view the OSS approach to AGI on balance – safer or less safe than the alternatives, and why? And how confident are you of your views on this?

Linas

Here, we must shift our focus from "what AGI is, today", to "what we believe AGI might be, in the future".

First of all, let's be clear: As history shows, the "bad guys" are always better funded, more organized, and more powerful than Robin Hood and his band of Merry Men. Were Robin Hood to create something of material value, the bad guys would find some way of getting their fingers on it; that's just how the world works. We could be talking about criminal gangs, corrupt cops, greedy corporations, rogue foreign governments, or, perhaps scariest of all, the paranoia of spy agencies out of control. The paranoia is justified; the Soviets used wire-tapping to keep the population under control, the Nazis used networks of informants, the current Iranian government uses the Republican Guard quite effectively, and J. Edgar Hoover exercised some extraordinary powers in the US. Let's not mention McCarthy and the Red Menace. It happens all the time, and it could happen anywhere, and it could happen in the US or in Europe. Some have argued that steps down this slippery slope have already been taken; few sleep better at night because of Homeland Security.

Let me be clear: I really don't think the "bad guys" are going to be some terrorists who use open-source AGI tech to unleash a holocaust against humanity. If there's one thing I do trust Homeland Security for, that is to catch the terrorists. They may be expensive and wasteful and coldly unfeeling, but they'll (probably) catch the terrorists.

Even in the most optimistic of circumstances, one might imagine a "benevolent" capitalistic corporation developing some advanced technology. Capitalism being what it is, this means "making money", even if it has deleterious effects on the environment, on employment, or, even "benignly", on customer satisfaction. Apple totally and completely dictates what can be found on the Apple app store: You cannot sue, you cannot appeal, you cannot in any way force them to open it up. There is no case law to stand on. You have no rights, in the app store.

Apple is not alone: Facebook has leaked any amount of information, Hotmail accounts have disappeared as if they've never existed, Amazon has deleted books from Kindles even though the customer had already paid for, and "owned" them. AT&T has silently and willingly gone along with warrantless wiretaps. Are you sure you want to entrust the future of AGI to players such as these?

Ben

Obviously I don't, which is one of the many reasons I'm enthused about OSS as a modality for AGI development.

Linas

And the intersection of AGI with the modern economy may have some much more dramatic consequences than these. The scariest thing about AGI, for me, is not the "hard takeoff" scenario, or the "bad guys" scenario, but is the economic dislocation. Is it possible that, before true AGI, we will have industrial quasi-AGI reshaping the economy to the point where it fractures and breaks?

There's an old, old worry that a "robot will take my job". In American folklore, John Henry "the Steeldriver" lost to a steam-powered drill in the 1840s, and it's been happening ever since. However, the industrialization of AGI is arguably of a magnitude greater than ever before, and massive economic dislocation and unemployment is a real possibility. Basically, I believe that long before we have "true AGI", we will have quasi-AGI, smart but lobotomized, feeling-less and psychotic, performing all sorts of menial, and not so menial, industrial tasks. If a robot can do it for less than $5/hour, a human will lose that job. When robots get good at everything, everyone will lose their job. They don't have to be "classic AGI" to be there.

What makes this a tragedy is not the AGI, but a political system that is utterly unprepared for this eventuality. The current rhetoric about welfare, taxes and social security, and the current legal system regarding private property and wealth, is I believe, dangerously inflexible with regards to future where an industrialized, lobotomized almost-but-not-quite-AGI robot can do any job better, and cheaper, than a human, and the only beneficiaries are the owners and the CEO's. Times are already tough if you don't have a college education. They'll get tough even for those who do. A 10% unemployment rate is bad. A 50% unemployment rate will result in bloodshed, civil war, and such political revolt that democracy, and rule of law, as we currently know it in America, will cease to exist. Optimistically, everyone would be on the dole. Realistically, all of the previously charted paths from here to there have lead through communism or dictatorships or both, and I fear both these paths.

Ben

Certainly, where these practical economic implications are concerned, timing means a lot. If it's 3 years from the broad advent of "quasi-AGI" programs that relieve 80% of humans from their jobs until the advent of truly human-level AGI, that's one thing. If it's 10 years, that's another thing. If it's 30 years, that's another thing.

But I wonder how OSS will play into all this. One obvious question is whether the potential for economic dislocation is any more or less if the first very powerful quasi-AGIs are developed via an OSS project rather than, say, a proprietary project of a major corporation? Or will the outcome will be essentially the same one way or another?

Conceptually, one could make a case that things may go better if the people are more fully in charge of the process of their own gradual obsolescence-as-workers. But of course, this might require more than just an open-source quasi-AGI framework, it might require a more broadly open and democratic society than now exists. This brings to mind RU Sirius's intriguing notion of an Open Source Party[36], bringing some of the dynamics and

---

[36] http://hplusmagazine.com/2011/02/24/open-source-party-2-0-liberty-democracy-transparency/

concepts of OSS and also of initiatives like Wikileaks to mainstream politics.

Plenty to think about, indeed!

Joel Pitt:
The Benefits of Open Source for AGI

Interview first published May 2011

Joel Pitt is another of my valued colleagues on the OpenCog open source AGI project. I first worked with Joel in 2001 when I helped advise his Master's thesis project applying genetic programming to biological data analysis. Later he joined the OpenCog project, and for a while served as leader of Hong Kong Polytechnic University's project applying OpenCog to create intelligent game characters. While we each have our individual orientations, through long years of dialogue we have probably come to understand the relevant issues considerably more similarly than a random pair of AGI researchers. So this interview is best viewed as a (hopefully) clear presentation of one sort of perspective on the intersection between open source and AGI.

Joel has a broad background in science and engineering, with contributions in multiple areas beyond AI, including bioinformatics, ecology and entomology. His PhD work, at Lincoln University in New Zealand, involved the development of a spatially explicit stochastic simulation model to investigate the spread of invasive species across variable landscapes. He is also a former Board member of Humanity+, and co-founded a startup company (NetEmpathy) in the area of automated text sentiment classification. In 2010, he won the student of the year award offered by the Canadian Singularity Institute for Artificial Intelligence.

Ben

What are the benefits you see of the open-source methodology for an AGI project, in terms of effectively achieving the goal of AGI at the human level and ultimately beyond? How would you compare it to traditional closed-source commercial methodology; or to a typical university research project in which software code isn't cleaned up and architected in a manner conducive to collaborative development by a broad group of people?

Joel

I believe open source software is beneficial for AGI development for a number of reasons.

Making an AGI project OSS gives the effort persistence and allows some coherence in an otherwise fragmented research community.

Everyone has their own pet theory of AGI, and providing a shared platform with which to test these theories I think invites collaboration. Even if the architecture of a project doesn't fit a particular theory, learning that fact is something that is valuable to know, along with where the approaches diverge.

More than one commercial project with AGI-like goals has run into funding problems. If the company then dissolves there will often be restrictions on how the code can be used or it may even be shut away in a vault and never be seen again. Making a project OSS means that funding may come and go, but the project will continue to make incremental progress.

OSS also prompts researchers to apply effective software engineering practices. Code developed for research often can end up a mess due to being worked on by a single developer without peer review. I was guilty of this in the past, but working and collaborating with a team means I have to comment my code and make it understandable to others. Because my efforts are visible to the rest of the world there is more incentive to design and test properly instead of just doing enough to get results and publish a paper.

Ben

How would you say the OpenCog project has benefitted specifically from its status as an OSS project so far?

Joel

I think OpenCog has benefited in all the ways I've described above.

We're fortunate to also have had Google sponsor our project for the Summer of Code in 2008 and 2009. This initiative brought in new contributors as well as helped us improve documentation and guides to make OpenCog more approachable for newcomers. As one might imagine, there is a steep learning curve to learning the ins and outs of an AGI framework!

Ben

In what ways would you say an AGI project differs from a typical OSS project? Does this make operating OpenCog significantly different from operating the average OSS project?

Joel

One of the most challenging aspects of building an OSS AGI project, compared to many other OSS projects, is that most OSS projects have a clear and simple end use. A music player plays music, a web server serves web pages, and a statistical library provides implementations of statistical functions.

An AGI on the other hand doesn't really reach its end use until it's complete, which may be a long time from project inception. Thus the rhythm of creating packaged releases, and the traditional development cycle, are less well defined. We are working to improve this with projects that are applying OpenCog to game characters and other domains, but the core framework is still a mystery to most people. It takes a certain level of time investment before you can see how you might apply OpenCog in your applications.

However, a number of projects associated with OpenCog have made packaged releases. RelEx, the NLP relationship extractor, and MOSES, a probabilistic genetic programming system, are both standalone tools spun out from OpenCog. We tentatively plan for an OpenCog 1.0 release sometime around the end of

2012. That will be an exciting milestone, even though it's a long way from our ultimate goals of AGI at the human level and beyond. That's part of our broader roadmap, which sketches a path from the current state of development, all the way to our end goal of self-improving human-level AGI, which we hope may be achievable by the early 2020s.

Ben

Some people have expressed worries about the implications of OSS development for AGI ethics in the long term. After all, if the code for the AGI is out there, then it's out there for everyone, including bad guys. On the other hand, in an OSS project there are also generally going to be a lot more people paying attention to the code to spot problems. How do you view the OSS approach to AGI on balance? Is it more safe or less safe than the alternatives, and why? And how confident are you of your views on this?

Joel

I believe that the concerns of OSS development of AGI are exaggerated.

We are still in the infancy of AGI development, and scaremongering by saying that it's too dangerous, and that any such efforts shouldn't be allowed to happen, won't solve anything. Much like prohibition, making something illegal or refusing to do it will just leave it to more unscrupulous types.

I'm also completely against the idea of a group of elites developing AGI behind closed doors. Why should I trust self-appointed guardians of humanity? This technique is often used by the less pleasant rulers of modern-day societies: "Trust us – everything will be okay! Your fate is in our hands. We know better."

Another advantage is that the open-source development process allows developers to catch one another's coding mistakes. When an OSS project reaches fruition, it typically has had many contributors and many eyes on the code to catch what smaller teams may not have found. An open system also allows for other Friendly AI theorists to inspect the mechanism behind an AGI system and make specific comments about the ways in which Unfriendliness could occur. When everyone's AGI system is created behind closed doors, these sorts of specific comments cannot be made, nor proven to be correct.

Further, to a great extent, the trajectory of an AGI system will be dependent on the initial conditions. Even the apparent intelligence of the system may be influenced by whether it has the right environment and whether it's bootstrapped with knowledge about the world. Just like having an ultra-intelligent brain sitting in a jar with no external stimulus will be next to useless, so will a seed AI that doesn't have a meaningful connection to the world. (Despite claims to the contrary, I can't see a seed AI developing in a ungrounded null-space.)

I'm not 100% confident of this, but I'm a rational optimist. Much as I'm a fan of open governance, I feel the fate of our future should also be open.

Ben
So what is our future fate, in your view?

Joel
That's a good question, isn't it. "When will the singularity occur?"... Would be the typical question the press would ask so that they can make bold claims about the future. But my answer to that is NaN. :-)

(For the non-programmer reader: NaN means "Not a Number", a commonly obtained error message.)

Ben
You reckon we just don't have enough knowledge to make meaningful estimates, so we just need to be open to whatever

happens and keep moving forward in what seems like a positive way?

Joel

Indeed!

Randal Koene:
Substrate-Independent Minds

Interview first published August 2011

Dr. Randal Koene [37] is a cross-disciplinary scientist whose background spans computational neuroscience, psychology, information theory, electrical engineering and physics. He currently leads the non-profit Carbon Copies, devoted to promoting and advancing mind uploading technology, and also the for-profit firm NeuraLink Co. He is also the Science Director of Dmitry Itskov's Global Future 2045 initiative. Previously he served as Director of Analysis at Halcyon Molecular [38], working on breakthrough methods in DNA sequencing and other projects, as Director of the Department of Neuroengineering at Tecnalia, the third largest private research organization in Europe, and as professor at the Center for Memory and Brain of Boston University.

Over the last decade, Randal has been perhaps the world's most vocal, steadfast and successful advocate of the idea of "mind uploading" or, as he now prefers to call it "substrate-independent minds." Putting a human brain in a digital computer, for example – or in general making practical technology for treating minds as patterns of organization and dynamics, rather than being tied to particular physical implementations. In 2011, when I edited a Special Issue of the *Journal of Machine Consciousness* on the topic of Mind Uploading, Randal was instrumental in helping me recruit quality authors for articles for the issue. His websites minduploading.org [39] and carboncopies.org [40] contain valuable information for anyone interested in these subjects – and in this

---

[37] http://rak.minduploading.org/
[38] http://halcyonmolecular.com/
[39] http://minduploading.org/
[40] http://carboncopies.org/

interview he gives a wonderful overview of the current state and future prospects of R&D regarding substrate-independent minds.

Ben

First, just to get the question out of the way, why the shift in terminology from "mind uploading" to "advancing substrate independent minds" (ASIM)? Is it just the same idea in new clothing or is there a conceptual shift also?

Randal

There are multiple reasons for the shift in terminology. The first reason was a confusion, at least on the part of the uninitiated, about what mind uploading means. There's uploading, downloading, offloading. With respect to memory, I've heard all three used before. But what's worse, uploading is normally associated only with taking some data and putting it somewhere else for storage. It does not imply that you do anything else with it, while the most interesting part about the objective is not just to copy data, but to re-implement the processes. The goal is to emulate the operations that have to be carried out to have a functioning mind.

The second important reason was that mind uploading does not clearly say anything about our objective. Mind uploading is a process, and that is exactly what it should stand for and how we use that term: Mind uploading is the process of transfer, a process by which that which constitutes a specific mind is transferred from one substrate (e.g. the biological brain) to another (e.g. a silicon brain).

There was also a slight conceptual shift involved in the adoption of the new Substrate-Independent Minds terminology. Or, if there was not a conceptual shift then the perceived shift can be blamed even more squarely on the original terminology.)

Ben

Elaborate a bit?

Randal

Consider this: When you have accomplished a transfer, i.e. a mind upload, then that mind is no longer in its original substrate. The mind could be operating on any sufficiently powerful computational substrate. In that sense, it has become substrate-independent. You managed to gain complete access and to transfer all of its relevant data. That is why we call such minds Substrate-Independent Minds. Of course, they still depend on a substrate to run, but you can switch substrates. So what do you do with a substrate-independent mind? Our interests lie not just in self-preservation or life-extension. We are especially interested in enhancement, competitiveness, and adaptability. I tried to describe that in my article *Pattern survival versus Gene survival,* which was published on Kurzweil AI.

Furthermore, we are not only interested in minds that are literal copies of a human mind, but have a strong interest in man-machine merger. That means, other-than-human substrate-independent minds are also topics of SIM.

Ben

Ah, for instance AGI minds that are created from the get-go to be substrate independent, rather than being uploaded from some originally substrate-dependent form... Yes, I've got plans to create some of those!

Randal

Right. Admittedly though, the current emphasis is heavily tilted towards the initial challenge of achieving sufficient access to the biological human mind.

We have been using the term ASIM on the carboncopies.org web-site, because we wanted to make it very clear that the organization is action-oriented and not a discussion parlor.

Ben

The "A" is for "Advancing".

Randal

And SIM is the usual abbreviation of Substrate-Independent Minds.

Ben

So that's a very, very broad concept – encompassing some current neuroscience research and also some pretty far-out thinking.

Randal

Indeed, there are many suggestions about how SIM might be accomplished. Those include approaches that depend on brain-machine interfaces, Terasem-like personality capture, etc. I imagine that most of those approaches can be quite useful, and many will bump into a number of shared challenges that any route to SIM has to overcome. But many approaches are also mired in philosophical uncertainty. For example, if you did live with a BMI for decades and relied more and more heavily on machine parts, would it be fair to say that you lose hardly anything when the biological brain at its origin ceased to function? That is the hope of some, but it seems a little bit like saying that society depends so much on machines that if all the people disappeared it would not matter. Similarly, if we capture my personality through careful description and recording of video and audio, at which point could I truly say that I had been transferred to another substrate? How does such an approach differ substantially from an autobiography?

Given those philosophical uncertainties, nearly everyone who is interested in SIM and has been working seriously in the field for a while is presently investigating a much more conservative approach. Those people include Ken Hayworth, Peter Passaro, but also well-known project leads who have an interest in emulating the functions of brain circuitry, such as Henry Markram, Jeff Lichtman and Ed Boyden. The most conservative approach is a faithful re-implementation of a large amount of detail that resides in the neuroanatomy and neurophysiology of the brain. The questions encountered there are largely about how much of the brain you need to include, how great the resolution of the re-implementation needs to be (neural ensembles, spiking neurons, morphologically correct neurons, molecular processes), and how you can acquire all that data at such scale and resolution from a biological brain? That conservative implementation of SIM is what I called Whole Brain Emulation many years ago. That term stuck, though it is sometimes abbreviated to "brain emulation" when the scope of a project is not quite as ambitious.

(**Editor's note:** *See the Whole Brain Emulation Roadmap*[41] *produced at Oxford University in 2007, based on a meeting involving Randal Koene and others.*)

Ben

So... in the vein of the relatively near-term and practical, what developments in the last 10 years do you think have moved us substantially closer to achieving substrate-independent human minds?

Randal

There are five main developments in the last decade that I would consider very significant in terms of making substrate-independent minds a feasible project.

The first is a development with its own independent drivers, namely advances in computing hardware, first in terms of memory and processor speeds, and now increasingly in terms of parallel computing capabilities. Parallel computation is a natural fit to neural computation. As such, it is essential both for the acquisition and analysis of data from the brain, as well as for the re-implementation of functions of mind. The natural platform for the implementation of a whole brain emulation is a parallel one, perhaps even a neuromorphic computing platform.

---

[41]
```
http://www.fhi.ox.ac.uk/selected_outputs/fohi_publ
ications/brain_emulation_roadmap
```

The second major development is advances in large-scale neuroinformatics. That is, computational neuroscience with a focus on increasing levels of modeling detail and increasing scale of modeled structures and networks. There have been natural developments there, driven by individual projects (e.g. the Blue Brain Project), but also organized advances such as those spearheaded by the INCF. Any project towards substrate-independent minds will obviously depend on a large and rigorous means of representation and implementation with a strong resemblance to modeling in computational neuroscience.

Third, actual recording from the brain is finally beginning to address the dual problems of scale and high resolution. We see this both in the move towards ever larger numbers of recording electrodes (see for example Ed Boyden's work on arrays with tens of thousands of recording channels), and in the development of radically novel means of access during the last decade. A much celebrated radical development is optogenetics, the ability to introduce new channels into the synapses of specific cell types, so that those neurons can be excited or inhibited by different wavelengths of light stimulation. It is technology such as this, which combines electro-optical technology and biological innovation that looks likely to make similar inroads on the large-scale high-resolution neural recording side. Artificial or synthetic biology may be the avenues that first take us towards feasible nanotechnology, which has long been hailed as the ultimate route to mind uploading and substrate-independent minds (e.g. recall the so-called Moravec procedure)

A fourth important type of development has been in the form of projects that are aimed specifically at accomplishing the conservative route to SIM that is known as whole brain emulation. There we see tool development projects, such as at least three prototype versions of the Automated Lathe

Ultramicrotome [42] (ATLUM) developed at the Lichtman lab at Harvard University. The ATLUM exists solely because Ken Hayworth[43] wants to use it to acquire the full neuronanatomy, the complete connectome of individual brains at 5nm resolution for reconstruction into a whole brain emulation. There are a hand full of projects of this kind, driven by individual researchers who are part of our network that actively pursues whole brain emulation.

The fifth development during the last decade is somewhat different, but very important. It is a conceptual shift in thinking about substrate-independent minds, whole brain emulation and mind uploading. Ten years ago, I could not have visited leading mainstream researchers in neuroscience, neural engineering, computer science, nanotechnology and related fields to discuss projects in brain emulation. It was beyond the scope of reasonable scientific endeavor, the domain of science fiction. This is no longer true.

Ben

Yes, that's an acute observation. I've noticed the same with AGI, as you know. The AI field started out 50 years ago focused on human-level thinking machines with learning and self-organization and all that great stuff, but when I got my PhD in the late 80s, I didn't even consider AI as a discipline, because it was so totally clear the AI field was interested only in narrow problem-solving or rigid logic-based or rule-based approaches. I was interested in AGI more than pure mathematics, but I chose to study math anyway, figuring this would give me background knowledge I would be able to use in my own work on AGI later on. Even 10 years ago it was hard to discuss human-level AI – let alone superhuman thinking machines – at a professional AI conference without someone laughing at you. Then 5 to 8 years ago, occasional brief workshops or special tracks relevant to

---

[42] http://www.mcb.harvard.edu/lichtman/ATLUM/ATLUM_web.htm

[43] http://geon.usc.edu/~ken/

human-level AI or AGI started to turn up at AAAI and IEEE conferences. And then I started the AGI conference series in 2006, and it's flourished pretty well, drawing in not only young researchers but also a lot of old-timers who were interested in AGI all along but afraid to admit it for fear of career suicide.

Randal

Yes, the situations are fairly similar – and related, because AGI and SIM share a lot of conceptual ground.

Ben

And I guess the situations are correlated on a broader level, also. As science and technology accelerate, more and more people – in and out of science – are getting used to the notion that ideas traditionally considered science fiction are gradually becoming realities. But academia shifts slowly. The "academic freedom" provided by the tenure system is counterbalanced, to an extent, by the overwhelming conservatism of so many academic fields, and the tendency of academics to converge on particular perspectives and ostracize those who diverge too far. Of course, a lot of incredibly wonderful work comes out of academia nonetheless, and I'm grateful it exists – but academia has a certain tendency toward conservatism that has to be pushed back against.

Randal

In fact I'm still mildly wary to list the names of ASIM-supportive scientists in publications, though I can if pressed. It can still be a slightly risky career move to associate oneself with such ideas, especially for a young researcher. But even so, it is now evident that a significant number of leading scientists, lab PIs, and researchers celebrated in their fields do see brain emulation and even substrate-independent minds are real and feasible goals for research and technology development. I can honestly say that this year, I am in such a conversation with specific and potentially collaborative aims and within the context of SIM on a bi-weekly basis. That development is quite novel and very promising.

Ben

So, let's name some names. Who do you think are some current researchers doing really interesting work pushing directly toward substrate-independent human minds?

Randal

I will try to keep this relatively short, so only one or two sentences for each:

- Ken Hayworth and Jeff Lichtman (Harvard) are the guiding forces behind the development of the ATLUM, and of course Jeff also has developed the useful Brainbow technique.

- Winfried Denk (Max-Planck) and Sebastian Seung (MIT) popularized the search for the human connectome and continue to push its acquisition, representation and simulations based on reconstructions forward, including recent publications in Science.

- Ed Boyden (MIT) is one of the pioneers of optogenetics, a driver of tool development in neural engineering, including novel recording arrays and a strong proponent of brain emulation.

- George Church (Harvard), previously best known for his work in genomics, has entered the field of brain science with a keen interest in developing high-resolution large-scale neural recording and interfacing technology. Based on recent conversation, it is my belief that he and his lab will soon become important innovators in the field.

- Peter Passaro (Sussex) is a driven researcher with the personal goal to achieve whole brain emulation. He is doing so by developing means for functional recording and representation that are in influenced by the work of Chris Eliasmith (Waterloo).

- Yoonsuck Choe (Texas A&M) and Todd Huffman (3Scan) continue to improve the Knife-Edge Scanning Microscope[44]

---

[44] http://research.cs.tamu.edu/bnl/kesm.html

(KESM), which was developed by the late Bruce McCormick with the specific aim of acquiring structural data from whole brains. The technology operates at a lower resolution than the ATLUM, but is presently able to handle acquisition at the scale of a whole embedded mouse brain.

- Henry Markram (EPFL) has publicly stated his aim of constructing a functional simulation of a whole cortex, using his Blue Brain approach that is based on statistical reconstruction based on data obtained from studies conducted in many different (mostly rat) brains. Without a tool such as the ATLUM, the Blue Brain Project will not develop a whole brain emulation in the truest sense, but the representational capabilities, functional verification and functional simulations that the project produces can be valuable contributions towards substrate-independent minds.

- Ted Berger (USC) is the first to develop a cognitive neural prosthetic. His prosthetic hippocampal CA3 replacement is small and has many limitations, but the work forces researchers to confront the actual challenges of functional interfacing within core circuitry of the brain.

- David Dalrymple (MIT/Harvard) is commencing a project to reconstruct the functional and subject-specific neural networks of the nematode C. Elegans. He is doing so to test a very specific hypothesis relevant to SIM, namely whether data acquisition and reimplementation can be successful without needing to go to the molecular level.

Ben

Well, great, thanks. Following those links should keep our readers busy for a while!

Following up on that, I wonder if you could say something about what you think are the most important research directions for SIM and WBE right now? What do the fields really need, to move forward quickly and compellingly?

Randal

Overall, I think that what SIM needs the most at this time are

1. to show convincingly that SIM is fundamentally possible
2. to create the tools that will make SIM possible and feasible
3. to have granular steps towards SIM that are themselves interesting and profitable for entrepreneurial and competitive effort

The first point is being addressed, as noted above. Points 2 and 3 are closely related, because the tools that we need to achieve SIM can be built and improved upon in a manner where their successive capabilities enable valuable innovations. Carboncopies.org will be organizing a workshop aimed specifically at that question in the fall.

I think that technologies able to represent and carry out in parallel the functions needed by a substrate-independent mind are highly valuable and may require improvements in neuromorphic hardware or beyond. And yet, my main concerns at this stage are at the level of acquisition, the analysis of the brain. Any technology that leads to greater acquisition at large scale and high resolution will have enormous potential to lead us further toward substrate-independent minds. That is why I pointed out large scale recording arrays and mixed opto-electronic/bio approaches such as optogenetics above. If we were to attempt whole brain emulation at a scale beyond the nematode at this time, then a structure-only approach, with attempted mapping from structure to function, using the ATLUM would be the best bet.

Ben
Stepping back a bit from nitty gritty, I wonder what's your take on consciousness? It's a topic that often comes up in discussions of mind-uploading... er, ASIM. You know the classic question. If you copied your brain at the molecular level into a digital substrate, and the dynamics of the copy were very similar to the original, and the functional behavior of the copy were very similar

to the original – would you consider that the original genuinely was YOU? Would it have your consciousness, in the same sense that the YOU who wakes up in the morning has the same consciousness as the YOU who went to sleep the previous night? Or would it be more like your identical twin?

Randal

You are really asking three questions here. You are asking what I consider sufficient for consciousness to be an emergent property of the reimplemented mind. You are also asking if I would consider a reimplementation to be equal to myself, to be myself. And you are asking if this would be me in specific circumstances, for example if there were multiple copies (e.g. an original brain and a copy) that existed and experienced things at the same time. I will try to answer each of these questions.

I personally believe that consciousness is not a binary thing, but a measure. It is like asking if something is hot. Hot relative to what? How hot? There should be measurable degrees of consciousness, at least in relative terms. I believe that consciousness is a property of a certain arrangement of a thinking entity, which is able to be aware to some degree of its own place in its model of the world and the experiences therein. I do think that such consciousness is entirely emergent from the mechanisms that generate thought. If the necessary functions of mind are present then so is consciousness. Does this require molecular-level acquisition and reimplementation? At this stage, I don't know.

If a copy of my mind were instantiated as you described and it told me that it was self-aware, then I would tend to believe it as much as I would believe any other person. I would also be inclined to believe that said copy was to any outside observer and to the world at large as much an implementation of myself as my original biological implementation (assuming that it was not severely limited mentally or physically).

Is such a copy indeed me? Is it a continuation of myself to the extent where a loss of my original biological self would imply no significant loss at all? That is where things become philosophically tricky. Those who argue that personal identity and the sense of self-continuity are illusions would probably argue that the means by which a copy was created and whether that copy was the only thing to continue to exist are irrelevant. They would be satisfied that self-continuity was achieved as much in such a case as it is when we wake up in the morning, or even from moment to moment. On the other end of the spectrum, you have those who would argue, even in a quantized universe, that the arrangement of instantiations from one quantum to its adjacent quantum states has some implication for real or perceived continuity. In that case, it can matter whether your brain was sliced and scanned by an ATLUM prior to reimplementation and reactivation, or if a transition was arranged through large-scale neural interfaces and a gradual replacement of the function of one neuron at a time. I have often described myself as a fence-sitter on this issue, and I still am. If you confront me with one of the two perspectives then I will argue the opposite one, as I see and personally sense relevance in both positions. If my stance in this matter does not change then it will have implications about the sort of uploading procedure that I would find cautiously satisfactory. In such a case, a procedure that assumes a problem of self-continuity would be the safe choice, as it would satisfy both philosophical stances.

Of course, even with such a procedure or in the event that there was no self-continuity problem then you can still end up with something akin to the identical twin situation. You could have two initially linked and synchronous implementations of a mind and identity. You could then sever the synchronizing connection and allow each to experience different events. The two would legitimately have been identical to begin with, but would gradually become more different. That is not really a problem, but rather an interesting possibility.

Ben

Yes, I tend to agree with your views.

But just to return to the gradual uploading scenario for a moment – say, where you move yourself from your brain to a digital substrate neuron by neuron. What do you think the process would feel like? Would it feel like gradually dying, or would you not feel anything at all; or some more subtle shift? Of course none of us really knows, but what's your gut feeling?

Randal

What would it feel like? Do you currently feel it when a bunch of your neurons die or when your neural pathways change? The gradual process is something we are very much accustomed to. You are not the same person that you were when you were 5 years old, but there is a sense of continuity that you are satisfied with. Our minds have a way of making sense out of any situation that they are confronted with, as for example in dreams. Assuming that a gradual process were sufficiently gradual and did not in effect seem like a series of traumatic brain injuries, I don't think that you would feel at all strange.

Ben

Well, I'm certainly curious to find out if you're right! Maybe in a couple decades.

Moving on – I wonder, what do you see as the intersection between AGI (my own main research area) and substrate-independent human minds, moving forward? Of course, once a human mind has been ported to a digital substrate, we'll be able to study that human mind and learn a lot about intelligent systems that way, including probably a lot about how to build AI systems varying on the human mind and perhaps more generally intelligent than the human mind. (At least, that's my view, please say so if you disagree!). But I'm wondering about your intuition regarding how general will be the lessons learned from digital versions of human minds. One point of view (let's call it A) could be that human minds are just one little specialized corner of mind-space, so that studying digital versions of human minds won't tell us that much about how to make vastly superhuman intelligence, or even about how to make non-humanlike (but roughly human level) general intelligences. Another point of view (let's call it B) could be that the human mind embodies the main secrets of general intelligence under limited computational resources, so that studying digital versions of human minds would basically tell us how general intelligence works. Do you lean toward A or B or some other view? Any thoughts in this general direction?

Randal
I think I lean a little bit towards A and a little bit towards B. I think that the human mind probably is very specialized, given that it evolved to deal with a very specialized set of circumstances. I look forward to being able to explore beyond those constraints. At the same time, I think that much of the most interesting work in AI so far has been inspired directly or indirectly by things we have discovered about thinking carried out in biological brains. I don't think we have reached the limits of that exploration or that exploring further within that domain would impose serious constraints on AGI. It is also true that much of what we think of when we think of AGI are in fact capabilities that are demonstrably within the bounds of what human intelligence is capable of. It seems that we would be quite happy to devise machines that are more easily able to interact with us in domains that we care about. So, from that point of view also, I see no problem with learning from the human mind when seeking to create artificial general intelligence. The existence of the human mind and its capabilities provides a reassuring ground-truth to AGI research.

Actually, I believe that there are many more areas of overlap between AGI and SIM research. In fact, they are close kin. It is not just that an AGI is a SIM and a SIM an AGI, but also that the steps needed to advance toward either include a lot of common ground. That effort, those steps required, will lead to insights, procedures, tools and spin-offs that impact both fields. The routes to AGI and to SIM are ones with many milestones. It is no coincidence that many of the same researchers who are active in one of the two fields are also active or strongly interested in the other.

Ben

Your latter point is an issue I'm unclear about, as an AGI researcher. I would suppose there are some approaches to AGI that are closely related to human cognition and neuroscience, and these have obvious close relatedness with WBE and human-oriented SIM. On the other hand, there may be other approaches to AGI owing relatively little to the human mind/brain – then these would relate to SIM broadly speaking, but not particularly to WBE or human-oriented SIM, right? So, partly, the relationship between AGI and SIM will depend on the specific trajectory AGI follows. My own approach to AGI is heavily based on human cognitive science and only loosely inspired by neuroscience, so I feel like its relatedness with WBE and human-oriented SIM is nontrivial but also not that strong. On the other hand, some folks like Demis Hassabis are pursuing AGI in a manner far more closely inspired by neuroscience, so there the intersection with WBE and human-oriented SIM is more marked.

Next, what about carboncopies.org? It seems an important and interesting effort, and I'm wondering how far you intend to push it. You've done workshops – do you envision a whole dedicated conference on ASIM at some point? Do you think ASIM is taken seriously enough that a lot of good neuroscientists and scientists from other allied fields would come to such a conference?

Randal

Carboncopies.org has four main reasons for existence:

1. To explain the fundamentals of SIM, demonstrating its scientific basis and feasibility.
2. To create and support a human network, especially those who can together bring about SIM. Among those, to facilitate multi-disciplinary for-profit and non-profit projects.

3. To create and maintain a roadmap, relying in part on access enabled through affiliation with Halcyon and such, but also relying in part on its independence as a non-profit.
4. To provide and maintain a public face of the field of SIM. That includes outreach activities such as publications and meetings.

SIM is obviously a multi-disciplinary field, so that one of the challenges is to carry out 1-4 across the spectrum, what our co-founder has called a polymath and initially technology agnostic approach.

We spent some time and effort on point 4 during the past year, but it looks like some of that activity may be well taken care of by others at present (e.g. Giulio Prisco and his teleXLR8 project). For this reason, our near future efforts concentrate more strongly on points 1 to 3. For example, we will ask a select group of experts to help us address the question of ventures and entrepreneurial opportunities in the area of SIM.

But yes, I do believe that at this point SIM has gained enough traction and feasibility that good scientists such as those mentioned above are interested in our efforts. We have confirmed the intention of several to participate in the expert sessions.

Ben

OK, next I have to ask the timing question. (I hate the question as a researcher, but as an interviewer it's inevitable.) When do you think we will see the first human mind embodied in a digital substrate? Ray Kurzweil places it around 2029, do you think that's realistic? If not what's your best guess?

And how much do you think progress toward this goal would be accelerated if the world were to put massive funding into the effort, say tens of billions of dollars reasonably intelligently deployed?

Randal
I work for Halcyon :)

Ben
A company that explicitly has WBE and SIM as part of its medium-term plan, although its short-term focus is mainly on revolutionizing DNA sequencing?

Randal
Correct. Therefore, I cannot give an unbiased estimate of the time to SIM or the acceleration that large-scale funding would provide. Achieving such funding is an explicit aim, and the acceleration should be such that SIM is achieved within our lifetimes. The reason why carboncopies.org exists is also to ensure a roadmap that lays the groundwork for several possible routes to SIM. The preferred routes are those that are achievable in a shorter period of time, and they will require the attraction and allocation of greater resources and effort. Perhaps 2029 is optimistic, but I would hope that SIM can be achieved not long after that.

Note that it is also a bit difficult to say exactly when SIM is achieved, unless SIM were achievable only by means of a Manhattan-style project with no significant intermediary results, until a human brain is finally uploaded into a SIM. I think that is an unlikely approach. It is much more likely that there will be stages and degrees of SIM, from a gradual increase in our dependence on brain enhancements to eventual total re-implementation.

Ben
Ah, that's a good point. So what do you think are some of the most important intermediate milestones on the path between here and a digitally embodied human mind? Are there any milestones so that, when they're reached, you'll personally feel confident that we're on the "golden path" technologically to achieving digitally-embodied human minds?

Randal

We are already on the golden path. :)

One of the most important milestones is to establish a well-funded organization that is serious about supporting projects that address challenges that must be overcome to make SIM possible. The inception of such organizations is now a fact, though their maturation to the point where the necessary support is made available is still a few years away.

Technologically, the most important milestones are

1. Tools that allow us to acquire brain-data a high resolution (at least individual neurons) and large scale (entire regions and multiple modal areas of a brain), and to do so functionally and structurally. There are smaller milestones as these tools are developed, and we will notice those as they enable valuable spin-off creations.

2. Platforms that enable brain-size processing at scales and rates at least equivalent to those required for the human experience. Again, the development of those platforms brings with it obvious signs of progress as their early versions can be used to implement previously impossible or impractical applications.

Applications of brain-machine interfaces straddle both development areas and may be indicators or milestones that demonstrate advances.

Ben

Hmmm.... Another question I feel I have to ask is: what about the quantum issue? A handful of scientists (Stu Hameroff[45] among the most vocal) believe digital computers won't be adequate to embody human minds, and that some sort of quantum computer (or maybe even something more exotic like a "quantum gravity computer") will be necessary. And of course

---

[45] http://www.quantumconsciousness.org/

there are real quantum computing technologies coming into play these days, like DWave's technology[46], with which I know you're quite familiar.

And there is some evidence that quantum coherence is involved in some biological processes relevant to some animals' neural processes, e.g. the magnetic field based navigation of birds[47]. What's your response to this? Will classical digital computers be enough? And if not, how much will that affect the creation of digitally embodied human minds, both the viability and the timeline?

Randal

I don't personally care much about debates that focus on digital computing platforms. I am not trying to prove a point about what digital computers can do. I am trying to make functions of the mind substrate-independent, i.e. to make it possible to move them to other processing substrates, whatever those may be.

Are there platforms that can compute functions of mind? Yes. We have one such platform between our ears.

Can other platforms carry out the computations? Can they carry out ALL of the processing in the same space and in the same time that it takes to carry them out in the original substrate? No.

I think it is probably impossible to do better than elementary particles already do at doing what they naturally do, and all that emerges from that. But do we care?

I would say that we do not.

By analogy, let us consider the emulation of one computing platform (a Macintosh) on another platform (a PC). Do we care if we can reproduce and emulate all of the aspects of the Mac

---

[46] http://dwavesys.com/
[47] http://www.technologyreview.com/blog/arxiv/23748/

platform, such as its precise pattern of electrical consumption, the manner and rate at which a specific Macbook heats up portions of its environment? We really don't. All we care about is whether the programs that we run on the Mac are also producing the same results when run on a Mac emulator on the PC. In the same sense, there are many levels at which the precise emulation of a brain is quite irrelevant to our intentions.

The interesting question is exactly where to draw the lines as we choose to re-implement structure and function above a certain level and to regard everything below that level as a black box for which specific transformations of input to output must be accomplished. What if some aspects do require quantum coherence? In that case one must decide if the particular feature that depends on quantum coherence is one that we care to re-implement, one that is essential to our intentions. If so, then a platform capable of the re-implementation will be needed. Would that change the time-line to SIM? Possibly, or possibly not if D-Wave makes rapid progress! ;)

Personally, I am quite interested in the possibilities that quantum computing may provide, even if it is unnecessary for a satisfactory SIM. That is, because I am interested in enhancements that take us beyond our current mental capabilities. It is nice to be able to use a quantum computer, but it may be even more interesting to experience quantum computation as an integral part of being, just as we experience neuromorphic parallel computation.

Ben

Thanks a lot Randal, for the wonderfully detailed answers. I look forward to my upload interviewing your upload sometime in the future.

João Pedro de Magalhães: Ending Aging

Interview first published April 2011

The quest to end aging and prolong human life indefinitely has stepped outside the worlds of religion and science fiction, and is now the stuff of serious science. The number of research biologists consciously and intently working toward this once-outrageous-sounding goal increases year after year. Exactly when we'll get to the end goal remains quite unclear – but the steady march of progress in that direction has become unmistakable.

In many cases, the work of life-extension-oriented scientists may be mostly funded by research grants or commercial projects aimed at curing specific age-associated diseases or other relatively down-to-earth goals. But even research approaching the matter of ending aging via this slightly indirect route can have dramatic impact on the core problem. And in some cases, it's hard to distinguish work on specific age-associated conditions from work aimed at ending the condition of aging itself.

As we peer closer and closer to the essence of aging, some researchers feel it looks less like a unified entity, and more like a combination of multiple complex phenomena in complex biological networks. Others feel there may be one or a handful of critical underlying biological processes driving the panoply of aging-associated phenomena. Either way, year by year we are gathering more and more data – and deeper and deeper ideas – pertinent to cracking the puzzle.

Since I started following the longevity research literature carefully back in 2002, one of my favorite researchers in the area has been João Pedro de Magalhães, a Portuguese biologist currently leading the Integrative Genomics of Aging Group[48] at the University of Liverpool. His work combines an evolutionary focus, a mastery of complex bioinformatics tools, an appreciation for complex biological networks, and a futurist mind-set. And his website senescence.info[49] has for many years been a valuable resource for both the novice and the experience longevity researcher.

João Pedro's personal web page[50] makes no secret of his ambitious perspective. "I'm a scientist and philosopher, a dreamer, and a transhumanist," he writes. And in keeping with the transhumanist ethos, he avers that "the human condition is only the beginning of the extraordinary journey of the mind through the universe. I defend humankind stands a better chance of success if we understand technology rather than try to ban it. Technologies like genetic engineering, cloning, cybernetics, and nanotechnology will allow us to escape our human limitations and evolve beyond our dreams. May the dreams of today become the future."

João Pedro also plays guitar, composes music and does stand-up comedy. But when I decided to interview him for H+ Magazine, I promised myself to resist the urge to ask him to participate in an online metal-fusion jam session, and instead resolved to probe into his views on the particulars of longevity research. I was especially interested to get his latest ideas about the relationship between development and aging, since this pertains to some of my own current work with Genescient[51], studying aging in long-lived fruit flies with a view toward discovering therapies for age-associated diseases in humans. I

---

[48] http://pcwww.liv.ac.uk/~aging/
[49] http://senescence.info/
[50] http://jp.senescence.info/
[51] http://genescient.com/

also wanted to get his take on some arguments[52] between aging researchers Aubrey de Grey and Michael Rose that I've been poking my nose into recently. I thought João Pedro would have some interesting comments, and I wasn't disappointed.

Ben

I'll start off with some general questions, and then we'll plunge into some nitty-gritty longevity biology a little later.

First of all I'm curious for your impressions of the overall trend of the life extension research field. You've been working on life-extension related biology for a number of years now, and I'm wondering if you've noticed any shift in peoples' attitudes toward this sort of work, during the course of your career? Do you think people are getting significantly more comfortable with the notion of radical life extension, or have we not reached that phase yet?

João Pedro

Clearly the biggest shift in attitudes since I started working in the field has come from Aubrey de Grey's SENS (Strategies for Engineering Negligible Senescence) program, which had a major impact in introducing the general public to radical life extension and I guess it also led to some soul searching in the research community. My impression is that most researchers working in the field are still skeptical and some even opposed to radical life extension. But SENS also forced many to discuss and even some to openly accept radical life extension as a long-term goal of the field.

Ben

Speaking of Aubrey – you're of course familiar with his notion of "Longevity Escape Velocity", aka "the Methuselarity." That is, to put it roughly: The point in time at which, if you're alive then, your odds of dying due to age-associated medical problems become

---

[52] http://hplusmagazine.com/editors-blog/michael-rose-aubrey-de-grey-contrasting-views-biological-immortality-humanity-caltech-c%5d

very low, due to ongoing and ever-improving biomedical advances. What's your take on the Methuselarity?

João Pedro

Methuselarity makes sense in theory, though if you look at the history of science what often happens is that, rather than small incremental improvements in a given field, there's usually one amazing breakthrough. I can see aging being cured by one amazing discovery – or triggered by a single breakthrough – rather than a continuous progress towards a cure.

Ben

And do you have any intuitive sense of when that breakthrough might occur? Tomorrow? 500 years from now? 15 years? 40?

João Pedro

Predicting when we will reach Methuselarity or the point of being able to cure aging is obviously incredibly hard. I used to be optimistic that it would happen around the midpoint of this century but have become more doubtful of it lately, in part because of funding difficulties. It's relatively difficult to obtain funding for life extension oriented work, and don't even dream about mentioning radical life extension in a grant application, compared to many other sorts of human biomedical research. I think it's clear that the more funding being directed at aging research, the more people working on it, the faster the progress, the higher the chances of a breakthrough.

The other critical issue is whether the technology to develop a cure for aging exists nowadays or whether progress in many adjacent fields, like nanotechnology, synthetic biology and stem cells, are a pre-requisite. I guess what we need to do to maximize our chances of curing aging is to invest more not only in basic aging research and biomedical sciences but also in several other key scientific areas, many of which can be seen as transhumanistic.

Ben

Yes, that certainly makes sense to me. And I'd add AI — my own main research field, though I've also done a lot of other stuff including longevity-related bioinformatics as you know — to your list.

So let me ask a question about that — about the hypothesis of advanced Artificial General Intelligences (AGIs) doing longevity biology themselves. I wonder how much do you think progress toward ending aging would be accelerated if we had an AGI system that was, let's say, roughly as generally intelligent as a great human scientist, but also had the capability to ingest the totality of biological datasets into its working memory and analyze them using a combination of humanlike creative thought and statistical and machine learning algorithms? Do you think with this sort of mind working on the problem, we could reach the Methuselarity in 5 or 10 years? Or do you think we're held back by factors that this amazing (but not godlike) level of intelligence couldn't dramatically ameliorate?

João Pedro

Based on your definition of AGI, I think this would truly advance progress in aging research (as well as in many other fields). As an observer of the AI field (but in no way an expert) my doubt, however, is whether an AGI system can be devised in a near future with such a combination of skills in terms of memory, creativity, intelligence, etc.

Answering your questions about the pace of progress, there are factors that not even a godlike intelligence could overcome, like the time it takes to do experiments on aging and government regulations on clinical testing. But if the progress becomes exponential due to AGI systems then I can see a cure for aging being developed in a matter of years.

Ben

OK... Well, it would be fun to debate the plausibility of achieving advanced AGI in the near term with you, but I guess that would take us too far afield! So now let's plunge into some of the biology of aging.

At the last Humanity+ conference[53] we had some great give-and-take between Michael Rose and Aubrey de Grey, and one of the issues they disagree on is the main cause of the aging phenomenon. To simplify a little, Michael Rose believes that antagonistic pleiotropy is the major cause of aging. Aubrey de Grey thinks it's accumulated damage. Which side do you come down on?

João Pedro

I think both perspectives are partly correct, though at different levels.

The antagonistic pleiotropy theory, the idea that genes with beneficial functions early in life may have detrimental actions late in life, explains aging from an evolutionary perspective. However, it doesn't explain the causes of aging at the mechanistic, molecular level, which I think is what people generally mean by "causes of aging".

As for accumulated damage, it can be considered a major cause of aging, yet "damage" is such a broadly defined term that I'm not satisfied with this concept. Almost any detrimental molecular or cellular event in the body can be defined as "damage", and many (perhaps most) forms of "damage" in the body do not cause aging. So what we need to find out is the specific molecular and cellular processes that drive aging, and unfortunately we don't know yet what these are. I believe that DNA damage accumulation, and its effects on cells and on the depletion of stem cell pools, could be a key cause of aging but this is only a guess.

Also, I'm keen on the idea – underrated by most experts – that developmental mechanisms contribute to aging.

---

[53] http://humanityplus.org/conferences/

Ben

Yes, I want to ask you about your work on the developmental approach to aging -- we'll get to that shortly! But first I want to ask another "Michael versus Aubrey question[54]"! A large part of the debate between Michael and Aubrey at the Humanity+ conference centered around "biological immortality" and the late-life phase. As you know, in some of his recent scientific works, Michael Rose has presented data and arguments in favor of a "late life" phase in fruit flies and other animals including humans. Basically, as I understand it, he's argued that during "late life" the odds of the organism dying during any given day (or year) becomes essentially constant. Others have argued this isn't a real phenomenon and it's an illusion caused by heterogenous aging patterns among different populations. And Aubrey tends to agree with this. What's your view?

João Pedro

Late life mortality plateaus have been observed in several populations of different species, including humans. My impression is that this is due to the heterogeneity intrinsic to any population. Even genetically identical organisms show phenotypic variation due to interactions with the environment and to sheer chance. Mathematical models also show that heterogeneity can explain mortality plateaus. Therefore, I think it's much more likely that heterogeneity causes mortality plateaus than some unknown biological phenomenon.

Ben

I see, so on that particular point you basically agree with Aubrey, it seems...

On the other hand, I have to admit Michael has pretty much convinced me that heterogeneity is not the sole explanation for the observed late-life plateau... He has a new book coming out

---

[54] http://hplusmagazine.com/editors-blog/michael-rose-aubrey-de-grey-contrasting-views-biological-immortality-humanity-caltech-c

soon that will present his mathematical arguments for this point in a lot of detail, so I'll be curious to get your reaction once his book comes out and you take a look at it!

João Pedro

I look forward to seeing what new arguments Michael has on the mortality plateaus debate.

Ben

OK, so let's get back to development. You've just mentioned that in your view, developmental processes are critical for understanding aging. Can you briefly elaborate on why? Is this basically an instance of antagonistic pleiotropy, or is there more to it?

João Pedro

The developmental theory of aging can indeed be seen as a form of antagonistic pleiotropy, of genes or mechanisms beneficial early in life (in this case for growth and development) being detrimental late in life and contributing to aging. The idea is very simple: Some of the same genetically-regulated processes that are crucial in development continue throughout adulthood and become harmful. For example, far-sightedness is thought to derive from the continual growth of eye lenses. During development the eye lenses must grow but it seems that the genetic program determining their growth continues in adulthood and this contributes to a specific age-related disease. My hypothesis is that several processes programmed in the genome for important roles during development become harmful late in life and contribute to aging. In an indirect sense, it's a form of programmed aging processes.

Ben

Yes, I see. But could you briefly list what some of these processes are, and how/why you hypothesize they become harmful?

João Pedro

The general concept is that changes occurring during early stages of life as part of programmed developmental processes then become damaging late in life. More work is necessary to establish exactly which processes fall into this category but there are already several examples.

For instance, a few years ago, Anders Sandberg and I developed a model of brain aging[55] which briefly argues that the continuation of the genetic programs that are essential for brain development early life contribute to cognitive aging and neurodegeneration. One example of a specific process is brain plasticity which necessarily decreases prior to adulthood, but as it continues to decrease in adulthood it will eventually become detrimental. There's also recent evidence that changes in gene expression during aging follow trajectories that are set during development and this is due to regulated processes, perhaps involving micro RNAs. One recent study found that the genetic program that coordinates growth deceleration during development persists into adulthood. In other words, most animals like humans are genetically programmed to stop growing and this involves molecular and cellular changes. But as the same molecular and cellular changes involved in growth deceleration persist into late life then these will be harmful.

Ben

Yes, that paper by you and Anders is very fascinating, and I think in the future it may be broadly viewed as a milestone toward the understanding of age-associated neurodegeneration. But of course it's a 2005 paper ... I wonder if any particularly exciting pieces of evidence in favor of that theory have arisen since then?

---

[55] www.liv.ac.uk/~aging/mad05.pdf

Some choice extracts from Cognitive aging as an extension of brain development[56] by João Pedro de Magalhães and Anders Sandberg:

> *After puberty, the priority is no longer adaptation or intellectual developmental. The brain must be more stable because the emphasis has shifted to reproduction and childbearing... Thus, synaptic plasticity, the activity-dependent modification of synapses, decreases with age, as does synaptic density.*
>
> *Brain plasticity and neuro-modulation continue to decrease even in aged individuals. "You can't teach an old dog new tricks", or so the proverb goes and, in fact, a decline in learning rate has been observed in elderly people as well as in numerous animal models. Brain plasticity in adulthood has also been shown to be decline. Therefore, our hypothesis is that, in later stages, this developmentally linked process continues – because there is no evolutionary pressure for it not to – and causes cognitive dysfunction. Our proposal is that the brain plasticity changes aimed at increasing the robustness of the human mind in childhood and adolescence later contribute to cognitive aging and eventually dementia.*

The following figure from the paper shows the number of neurons (black line) and synapses (gray line) during the lifespan:

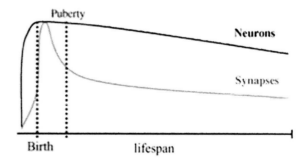

---

[56] www.liv.ac.uk/~aging/mad05.pdf

And, about the relation between calorie restriction, development and neurodegeneration:

In line with the delay of cognitive aging in CR mice, CR in rodents attenuates the age-related decline in neurotransmitters during aging as well as some neuronal receptors, and induces the expression of genes involved in brain plasticity. In fact, CR induces the expression of developmentally regulated genes in the mouse brain when compared to age-matched controls. The next step is linking these players, such as neurotransmitters, neurotrophic factors, and hormones, to the developmental program in order to understand the complex transcriptional control involved.

And about oxidative stress (ROS, Reactive Oxygen Species) and neurodegeneration and the complex networks there-involved:

Considerable evidence exists that ROS generation increases with age and oxidative stress is associated with learning impairments. Yet our proposal is that this increase in ROS with age is a consequence of the lack of responsiveness in the brain to ROS. ROS may then indirectly cause damage and so a feedback loop is formed between ROS trying to stimulate synaptic plasticity and oxidative stress. Therefore, our argument is that ROS do not trigger cognitive aging, but rather are under control by developmental mechanisms and thus form part of its signaling cascade. Like many other factors, ROS are deregulated with age due to the actions of the developmental program and thus cause damage.

João Pedro

There've been a few papers along these lines, but not that many. Arguably the best I've seen recently is one by a group of researchers in Shanghai, on "MicroRNA, mRNA, and protein expression link development and aging in human and macaque brain[57]".

The abstract of "MicroRNA, mRNA, and protein expression link development and aging in human and macaque brain," by Mehmet Somel and his colleagues at the Partner Institute for Computational Biology, at the Chinese Academy of Sciences in Shanghai:

> *Changes in gene expression levels determine differentiation of tissues involved in development and are associated with functional decline in aging. Although development is tightly regulated, the transition between development and aging, as well as regulation of post-developmental changes, are not well understood. Here, we measured messenger RNA (mRNA), microRNA (miRNA), and protein expression in the prefrontal cortex of humans and rhesus macaques over the species' life spans. We find that few gene expression changes are unique to aging. Instead, the vast majority of miRNA and gene expression changes that occur in aging represent reversals or extensions of developmental patterns. Surprisingly, many gene expression changes previously attributed to aging, such as down-regulation of neural genes, initiate in early childhood. Our results indicate that miRNA and transcription factors regulate not only developmental but also post-developmental expression changes, with a number of regulatory processes continuing throughout the entire life span. Differential evolutionary conservation of the corresponding genomic regions implies that these regulatory processes, although beneficial in development, might be detrimental in aging. These results suggest a direct link between developmental regulation and expression changes taking place in aging. [emphasis added]*

---

[57] http://www.ncbi.nlm.nih.gov/pubmed/20647238

Ben

Yes, that's outstanding stuff; and I confess I missed that paper when it came out, though Google reveals it was written up in Science Daily[58] not long ago! Thanks for bringing me up to speed. There's so much good biology published these days, it's hard to keep up with it all.

We'll be doing a Humanity+ conference in Hong Kong in November, and I'll be trying to pull in some Chinese researchers doing life extension work, as well as cognitive neuroscience and other areas. I'll definitely try to get someone from that lab to come talk about their work.

Also I'm thinking about the connections between these ideas and my own work applying AI bioinformatics to genetics data from Genescient's super-long-lived "Methuselah" fruit flies[59] that have been selectively bred to live four times as long as regular flies of the same species. It will be interesting to look for corroboration for the neuro-development neuro-degeneration connection in the Methuselah fly gene sequence data when we obtain it in a month or two.

Although, I suppose that to really explore that theory it would also be nice to have time-series gene expression data from Methuselah versus control flies over their whole lifespan, to see the perturbations in both the neurodevelopment and neurodegenerative phases.

João Pedro

Yes, I think you'd need data across the lifespan to study development-aging links.

---

[58] http://www.sciencedaily.com/releases/2010/07/100719174903.htm

[59] http://www.hplusmagazine.com/editors-blog/ais-superflies-and-path-immortality

Ben

Well, I'm sure Genescient will get that sort of data before too long. One of the wonderful things about doing biology in the era of accelerating change is that experimental data collection gets cheaper year upon year, at an amazing rate. This of course is one of the things driving the outstanding recent progress in biology.

So much data is gathered these days, that even using the best AI tools we have available, we struggle to deal with it, though it's a fun struggle! But of course, if we have a good theory of what kinds of patterns we're likely to see in the data, that makes the analysis process a bit easier. Which is part of why your own theories on aging interest me so much.

João Pedro

I should emphasize that this developmental theory of aging (or DevAge as I like to call it) is in contrast with most other theories of aging which tend to focus on random damage accumulation, often from by-products of metabolism.

Ben

Yes... In that respect your views seem closer to those of Michael than of Aubrey. But you focus on some fairly specific aspects of antagonistic pleiotropy, related to the development process.

João Pedro

Well, what DevAge suggests is that some aspects of aging follow pre-determined patterns encoded in the genome as part of developmental processes. In other words, the genome does indeed contain instructions that drive aging. For anti-aging interventions this has profound implications because it means that if we can unravel the regulatory elements of these processes via genomics and bioinformatics then we may be able to manipulate them and therefore develop new anti-aging interventions.

Ben

So, given this perspective, what research are you focusing on recently? And how does it fit into the big picture of combating aging and increasing human health span?

João Pedro

My strategy for fighting aging has always been to first learn more about the precise molecular, cellular and genetic causes of aging and then, with more specific targets in mind, develop interventions.

Ben

Yes, I see: Understand first, cure after. It's an appealing approach to me, as a scientist. On the other hand, we have to recall that none of the broadly successful vaccines were developed based on deep understanding – they all basically came out of educated guesswork and trial and error. One of Aubrey's ideas is that we may be able to extend life significantly via this sort of educated guesswork and tinkering, even without a full fundamental understanding of the underlying biological networks.

João Pedro

Breakthroughs based on fairly limited understanding are of course possible. But I must point out that aging is different beast than a pathogen. If what you aim to do is kill a virus, bacteria or even a cancer cell then you may not necessarily require a deep understanding of how your foe works. For instance, you can just screen for drugs that kill your target without killing human cells. Aging, however, affects multiple systems at different biological levels. I think we stand a better chance of retarding and ultimately curing aging if we better understand its basic mechanisms as this allows us to better focus our efforts. As such, our lab has been doing a lot of work at the genetic level and trying to identify genes that are important for longevity. This involves experimental work (including with next-generation sequencing technologies) and bioinformatics to predict from large datasets which genes are the important ones and even whether they could be suitable drug targets.

Another approach I've always been keen on is comparative biology or trying to identify the genes and processes determining species differences in aging (e.g., why humans live over 100 years while mice live only 4 years?). Among other studies, we've been doing a lot of work on the naked mole-rat, a long-lived and cancer-resistant mammal, with the goal of identifying genes that protect against aging and cancer.

Ben

Hmmm... Yes, that's a different approach from the ones most scientists are taking. Are there any specific, interesting conclusions that have come out of this comparative approach so far?

João Pedro

I have to say there's still a lot of work to be done before we have a coherent picture of why some species live longer than others. One thing we did show a few years back was that, contrary to many predictions, long-lived species do not have lower metabolic rates, so something else must be underlying their long lifespans. There's some results hinting that some repair pathways, like DNA repair, may be optimized in long-lived mammals but no conclusive results yet. The problem is that it's not easy from a practical perspective to study multiple species, including the long-lived ones like whales. This is one of the reasons why the genomics revolution is so important for comparative approaches to aging. The availability of a large number of genome sequences allows us to identify potential genes and mechanisms involved in the evolution of long lifespans using bioinformatics. Our lab is already working on this and as more genomes are sequenced our capacity and statistical power will increase exponentially and I suspect we'll start to have some good answers soon.

Ben

You've done some work related to caloric restriction too, is that right?

João Pedro

Yes, we've also been working on dietary manipulations of aging, such as caloric restriction, again to predict key genes that modulate the life-extending effects of CR and may be suitable targets for drug development.

Ben

I did some work like that myself, together with Lucio Coelho at Biomind. We used our AI tools to cross-analyze four gene expression datasets from mice under calorie restriction, and look for the genes that seemed most important across all the experiments. We came up with a lot of cool stuff that wasn't in any of the published literature. I know the sirtuins got a lot of attention as genes related to CR – and they did come out of our analysis, but they weren't at the top of the list. For instance, we got a lot of information pointing to a significant role for Mrpl12 – a nuclear gene coding for mitochondrial ribosomal proteins. I still don't know how Mrpl12 plays into the CR story. There's a lot unknown about the mechanisms of CR, and a lot that can be learned from them I'm sure.

João Pedro

For transhumanists, I guess CR may be of limited value as it only retards aging to some degree; but I think we need to take one step at a time. If we could retard aging using drugs that mimic effects of CR then this would bring enormous impetus to the field and allow us to then explore more ambitious avenues.

Ben

Seems we'll need to wrap up soon – we've covered a lot of ground! But I want to ask one more thing. Looking beyond your own work, of all the work currently going on in the anti-aging research field, what excites you most? What do you think has the most promise?

João Pedro
The finding that rapamycin can extend lifespan in middle-aged mice was exciting and arguably the most important in anti-aging research of the past 2-3 years. Because rapamycin has serious side-effects, however, it's not suitable as an anti-aging treatment, but this discovery could open the door for development of other less harmful drugs targeting the same pathway. There's also very exciting progress being done in adjacent areas that could have a great impact on anti-aging research, like regenerative medicine, synthetic biology and genomics. The recent progress in genomics has been amazing and the fact we'll soon be able to sequence virtually everyone's genome is tantalizing. The work on reprogramming of adult cells to become pluripotent stem cells has also been remarkable.

Ben
It's certainly an exciting time to be working in biology. So much amazing scientific progress, year after year – and then, such an ongoing stream of reminders as to how the (even more amazing) subtlety and complexity of biological networks repeatedly eludes our attempts at simplified understanding.

For instance, the fact that CR extends maximum lifespan is amazing, and it's great that we have the technology to isolate some of the underlying genetic factors like the sirtuins, Mrpl12, and so forth. But then Sirtris's attempts to commercially exploit our understanding of the sirtuin pathway for life extension pharmaceuticals, have been fairly problematic so far. And like you say, the rapamycin discovery was incredible, but side-effects due to various complex biological networks mean it's not practically usable in humans, though it gives some interesting directions for research.

On the one hand the complexity of living systems is incredibly intellectually interesting; and the other hand, it means that we (or our AI systems) may have to become pretty damn good at understanding complex self-organizing systems to have that Methuselarity. And yet, in spite of all this complexity (and of course powered by the complex biological networks in our brains and our society, and the complex digital information networks we've created), we *are* making progress, year on year.

Aubrey De Grey: Aging and AGI

Interview first published June 2009

No one has done more to bring the viability of ending aging to the attention of the general public and the scientific community than Aubrey de Grey. His specific plan for working toward this critical goal, which goes by the name of SENS (Strategies for Engineering Negligible Senescence), is currently the subject of research in a number of biology labs. His book Ending Aging[60] (co-authored with Michael Rae) is required reading for anyone interested in longevity or transhumanism; and his website sens.org[61] is also full of relevant information.

Aubrey and I have always had a particular rapport, due to our common interest in both AGI and life extension research. Aubrey was an AI researcher before he embarked on his current focus on life extension. However, we have taken significantly different approaches in our work, in both AI and biology – which means we always have a lot to talk about! In this brief and somewhat unsystematic dialogue, my goal was simply to gather a few of Aubrey's opinions on some longevity research issues touching my current interest. Hopefully they will touch your interest as well!

Ben

Jumping right into the middle of the discussion we were having at the Humanity+ @ Caltech conference: Your view seems to be that accumulated damage, rather than (as some others like Michael Rose would have it) antagonistic pleiotropy is the main culprit in human aging. Could you briefly summarize your reasons for this?

---

[60] http://www.amazon.com/Ending-Aging-Rejuvenation-Breakthroughs-Lifetime/dp/0312367066

[61] http://sens.org/

Aubrey

Well, I'm not sure that we can really say that accumulation of damage and antagonistic pleiotropy are alternatives. They are answers to different questions. Damage accumulation is a mechanistic hypothesis for how aging occurs, and AP is an evolutionary hypothesis for why it occurs. There are certainly some types of damage accumulation with aging that are caused as side-effects of machinery that is useful in early life - an example would be the accumulation of potentially toxic senescent cells that have arrested as a way to stop them from becoming cancerous - and that's basically all that the AP concept proposes. I don't think Michael thinks that aging is a maladaptive continuation of development, or some other "programmed" process. I think he agrees with me, and most other gerontologists, that aging is caused by damage accumulation. The only question is how that damage accumulation changes with age.

Ben

Hmmm... Well when I posed a similar question [62] to Joao Pedro deMagalhaes. He said:

> "Accumulated damage can be considered a major cause of aging, yet "damage" is such a broadly defined term that I'm not satisfied with this concept. Almost any detrimental molecular or cellular event in the body can be defined as "damage", and many (perhaps most) forms of "damage" in the body do not cause aging. So what we need to find out is what are the specific molecular and cellular processes that drive aging, and unfortunately we don't know yet what these are."

Aubrey

I don't agree with this. First, Joao Pedro is putting the cart before the horse in terms of the definition of "damage." Yes, sure, there

---

[62] http://www.liv.ac.uk/~aging/%5d

are lots of possible definitions, but I've always been clear about mine:

1. "damage" is a side-effect of metabolism
2. "damage" accumulates
3. once abundant enough, "damage" MAY contribute to age-related ill-health

Second, because of the third criterion above, I don't agree that we need to know for sure which accumulating side-effects of metabolism contribute to age-related ill-health, and which do not. I think we should just go after repairing all those that MIGHT so contribute. If we fix a few things we didn't need to fix, no big deal, whereas if we waste time on further analysis of which things to pursue and which not to, we delay the overall outcome.

Ben

Joao Pedro also says:

> "The developmental theory of aging (which I advocate) can be seen as a form of antagonistic pleiotropy, of genes or mechanisms beneficial early in life (in this case for growth and development) being detrimental late in life and contributing to aging. The idea is very simple; Some of the same genetically-regulated processes that are crucial in development continue throughout adulthood, where they then become harmful.
>
> For example, far-sightedness is thought to derive from the continual growth of eye lenses. During development the eye lenses must grow, but it seems that the genetic program determining their growth continues in adulthood and this contributes to a specific age-related disease. My hypothesis is that several processes programmed in the genome for important roles during development become harmful later in life, and contribute to aging. In an indirect sense, it's a form of programmed aging processes."

I've discussed Joao Pedro's perspective with Michael and he's largely sympathetic though he's not sure development-related antagonistic pleiotropy plays quite as large a role as Joao Pedro thinks, relative to other sorts of antagonistic pleiotropy.

Any reaction to these notions as articulated in the above paragraphs?

Aubrey

I agree with Michael that the few such lifelong processes that exist are peripheral to aging. I don't even agree with JP's example: Actually, far-sightedness is mainly caused by glycation-derived crosslinking.

However, I don't know what you/Michael may mean by "other sorts of antagonistic pleiotropy." Maybe you mean the sort I mentioned - accumulating damage from aspects of metabolism that exist to prevent the accumulation of other sorts of damage earlier in life, but maybe you don't.

Ben

Pursuing the same train of thought a little further, Michael Rose presents an argument in favor of a "late life" phase in fruit flies and other animals including humans. He argues that during "late life", the odds of the organism dying during any given year becomes essentially constant. However, this seems at odds with the idea that accumulating damage plays a large role in aging , because it seems that damage would just keep progressively accumulating as the organism gets older, rather than stopping to accumulate when "late life" occurs.) I know this is a technical matter, but it's an important one, so could you try your best to summarize your views on "late life" in a nontechnical way?

Aubrey

You've got it. I think it is highly implausible that an individual's accumulation of damage will ever cease to accelerate, let alone slow down or stop, after some age. It's not completely impossible that that could happen; That after some point the individual could, for example, shift to a different lifestyle (like a fly stopping flying) that would drastically reduce the rate of damage accumulation, but I don't buy it. Michael adopted this view as a result of what I have shown was a premature and oversimplistic mathematical analysis of the distribution of ages at death of his flies, an analysis that seemed to show that damage accumulation must indeed slow down. I've shown that the data are in fact entirely compatible with a lifelong acceleration in damage accumulation, given a modest degree of variation between individual flies in the initial rate of damage accumulation and (more importantly) the rate at which that rate accelerates via positive feedback. Thus, while I don't claim that Michael's interpretation can be excluded on the basis of existing data, I prefer the biologically more natural alternative that damage accumulation does indeed accelerate throughout life.

Ben

On a different note, I wonder how much do you think progress toward ending aging would be accelerated if we had an AGI system that was, let's say, roughly as generally intelligent as a great human scientist, but also had the capability to ingest the totality of biological datasets into its working memory and analyze them using a combination of humanlike creative thought and statistical and machine learning algorithms? Do you think with this sort of mind working on the problem, we could reach the Methuselarity in 5 or 10 years? Or do you think we're held back by factors that this amazing (but not godlike) level of intelligence couldn't dramatically ameliorate?

Aubrey

I think it's highly unlikely that such a system could solve aging that fast just by analyzing existing knowledge really well; I think it would need to be able to do experiments, to find things out that nobody knows yet. For example, it's pretty clear that we will need much more effective somatic gene therapy than currently exists, and I think that will need a lot of trial and error. However, I'm all for development of such a system for this purpose, as firstly, I might be wrong about the above, and secondly, even if it only hastens the Methuselarity by a small amount, that's still a lot of lives saved.

Ben

Yeah, I think that's probably right. But of course, a sufficiently advanced AGI system could also design new experiments and have human lab techs run them – or it could run the experiments itself using robotized lab equipment.

Aubrey

Right. I was explicitly excluding that scenario, because you seemed to be.

Ben

There have already been some simple experiments with completely robotized labs, where the experiments themselves are specified by AI algorithms, so there's a fully automated cycle from AI experiment specification, to robotized experiment execution, to ML data analysis, back to AI experiment specification, etc. But of course in the cases done so far, the experiments have been "templatized" rather than requiring creative, improvisatory experimental design.

Aubrey

Right.

Ben

I suppose the relevance of this sort of approach is relative to one's judgment of the relative timing of progress in AGI versus non-AGI-based life extension R&D. If you think we can make a Methuselarity by 2030 without AGI, but that developing AGI capable of providing dramatic help to life extension research will take till 2040, then AGI will be low priority for you (insofar as your goal is life extension). But if you think it will take till 2060 to make a Methuselarity without AGI, whereas AGI capable of dramatically helping life extension research could likely be created by 2030; then it will make sense for you to advocate a lot of resources going into AGI, even if your ultimate main focus is life extension.

Aubrey

Right. Except that of course AGI research is much cheaper, so there's really no reason to prioritise one over the other.

Ben

So it would appear to me, based on your comments, that you think the timeline to "Methuselarity without AGI help" is probably sooner than the timeline to advanced AGI that could dramatically accelerate life extension research". My QUESTION, at too long last, is: Is this a correct inference of mine, regarding your subjective estimates of these two timelines?

Aubrey

Not really, no. I honestly don't have a feel for how far away AGI of the sort that could really help is. It may be impossible (the Monica Anderson view). It may be rather easy (your view). It may be reasonably easy to create but really hard to make safe (Eliezer's view). I honestly haven't a clue.

Ben

Another factor that occurs to me, when discussing AGI and life-extension research, is the relative cost of the two types of research. If I'm right about AGI, the cost to achieve it could be tens of millions. Whereas non AGI based life extension research could very easily cost billions. So the question is: What's your current rough cost estimate for achieving Methuselarity? Are you still thinking of it as a multi-billion dollar endeavor? (Which of course is quite affordable by society, given the massive amount we spend on healthcare.)

Aubrey

Definitely multi-billion, yes; probably trillions, if we include all the medical delivery infrastructure and manpower. But the question then arises of whether AGI will cut the cost of those later stages as well as of the early stages. I don't really see why it shouldn't.

Between Ape and Artilect

David Brin: Sousveillance

Interview first published May 2011

Isaac Asimov's classic 1956 story "The Dead Past" describes a technology that lets everyone spy on everyone else everywhere. The government tries to keep this technology secret and arrest the scientists moving to publicize it, but some rebels let the cat out of the bag. A government official, once he realizes the technology has been made irrevocably public, utters the following lines to the scientists who released it:

> "Happy goldfish bowl to you, to me, to everyone, and may each of you fry in hell forever. Arrest rescinded."

Asimov's government official assumes that "everyone sees everyone else" is a nightmare scenario. But would it be, really? What if some version of Asimov's vision came true, and everyone could watch everyone else all the time?

This potentiality has been labeled "sousveillance" – and it may seem like a scary idea, since we're used to cultures and psychological habits predicated on greater levels of privacy. But this is exactly the future projected by a handful of savvy pundits, such as computer scientist Steve Mann and science fiction writer David Brin.

"Sous" and "sur" are French for "under" and "over" respectively. Hence, *sur*veillance is when the masters watch over the masses. *Sous*veillance is where everybody has the capability to watch over each other, peer-to-peer style – and not even the rulers are exempt from the universal collective eye. It's generally meant to imply that citizens have and exercise the power to look-back at the powers-that-be, or to "watch the watchmen."

Steve Mann conducted a series of practical experiments with video life-logging – recording everything he saw through his life, on video – which gave sousveillance a concrete and easily-understandable face. Following up early explorations in the 1980s, starting in 1994, Mann continuously transmitted his everyday life 24 hours a day, 7 days a week, attracting attention from media and futurist thought leaders. And Brin's 1998 *The Transparent Society* is a masterful nonfiction tome exploring the implications of sousveillance – and arguing for its benefits, as compared to the viable alternatives.

In his book, Brin makes a fairly strong argument that, as monitoring technology advances and becomes pervasive, these are the two most feasible options. Being monitored will become inescapable, it will be a matter of who is *able* to do the monitoring: Only a "trusted" central authority, or everyone in the society. Both have obvious disadvantages, but if persistent and pervasive monitoring is an inevitability, then sousveillance certainly bears consideration as an alternative to surveillance.

Like surveillance, sousveillance can occur to varying degrees, ranging from the ability of everyone to observe everything that goes on only in public places (as captured by omnipresent security cameras and other monitoring devices), to the ability for everyone to eavesdrop on everyone else's phone calls and personal communications. An argument in favor of more widespread sousveillance is that if the government is eavesdropping in these different ways, we may be better off if we all can do so as well, and therefore also eavesdrop on the government eavesdroppers. As Brin puts it, "You and I may not be able to stop the government from knowing everything about us. But if you and I know everything about the government, then they'll be limited in what they can DO to us. What they do matters more than what they know."

More radically futuristic options, where thoughts and feelings are made subject to observation via brain-monitoring technologies, form a common nightmare scenario in science fiction (though more commonly portrayed as surveillance, not sousveillance).

But even the more prosaic forms of sousveillance will have dramatic implications.

Brin himself tends not to view sousveillance as scary or disturbing, using the analogy of people sitting at different tables in a restaurant, who *could* eavesdrop on each others' conversations, but choose not to. Even nosy people generally refrain, because the eavesdropper is likely to be caught doing so, and snooping is disdained. He reckons that if sousveillance became a reality, new patterns of social tact would likely evolve, and society and psychology would self-organize into some new configuration, which would leave people significant privacy in practice, but would also contain the ever-present potential of active sousveillance as a deterrent to misdoings. This can be illustrated by extending the restaurant analogy; if universal sousveillance means that all peeping toms are always *caught in the act,* then such a society might wind up with more privacy than you'd expect.

Indeed, modest evidence for Brin's optimistic perspective exists already, in the shifting attitudes of the younger generation toward privacy. A significant percentage of young people seem not to care very much what others may know about them, openly putting all sorts of conventionally-considered-private information online. And in line with Brin's restaurant analogy, even though I could find out a lot of private information about a lot of people I know via their various online profiles, I very rarely bother to. And the psychological makeup of the younger generation does seem to be subtly but significantly shifted, due to this limited "online sousveillance" that has arisen. One may argue that society is slipping toward sousveillance bit by bit – implicitly and incrementally rather than in an explicitly discussed and deliberated way; as the Net comes to govern more and more of our lives, and personal information becomes more and more available online.

13 years after the publication of *The Transparent Society*, I was pleased to have the opportunity to pose David a couple

questions about the potential future implications of sousveillance. I found his confidence in the feasibility and potential value of sousveillance undiminished – but also detected a note of frustration at the ambiguity (and sometime skepticism or downright hostility) with which today's general public views Enlightenment ideals like calm reasonable analysis, which Brin views as important for nudging society toward positive uses of sousveillance technology.

Ben

What's your current thinking regarding the effects of widespread sousveillance on society? Let's say on a fairly open and democratic society like the US, How would our social structures change?

David

It depends somewhat on just how widespread the sousveillance is. "Truly pervasive sousveillance" could range all the way to radical levels portrayed in Damon Knight's story "I see you", in which the future is portrayed without any secrecy or even privacy at all. Knight shows a humanity that is universally omniscient -- any human, even children, can use a machine to view through any walls. And Knight depicts us adapting, Getting used to simply assuming that we are watched, all this time.

Now let me be clear, I do not care for this envisioned tomorrow! In *The Transparent Society* I devote a whole chapter to how essential some degree of privacy is for most people. I argue that in a society full of liberated people empowered with highly functional sousveillance technology, sovereign citizens, able to apply sousveillance toward any center of power than might oppress them, will likely use some of that sovereign power to negotiate with their neighbors, and come up with some way to leave each other alone.

This is the logical leap that too few people seem able to make, alas. That fully empowered citizens may decide *neither* to hand power over to a Big Brother; *nor* to turn into billions of oppressive little brothers. They might instead decide that the purpose of light is accountability. And shoving too much light into the faces of others, where accountability isn't needed, well, THAT would also be an abuse, a socially unacceptable activity. One that *you* can be held accountable for.

You can imagine a cultural adaptation to sousveillance, such that you can not only keep an eye on tyrants, but also catch peeping Toms; that is where we may get the best of both worlds. Both freedom and a little... Just a little... Privacy.

Ben

Yes, I can see the possibility of that. But even in this "best of both worlds" scenario, I guess sousveillance would have some real impact on how we think and live, right? What do you think would be the *psychological* effects of widespread sousveillance. How would it change our minds, our selves, our relationships? Do you think it would make us mentally healthier, or could it in some cases make us more neurotic or insane? Would some existing personality types vanish and others emerge? Would different sorts of collective intelligence and collaboration become possible?

David

Well, certainly people who are simmering in borderline insanity – stewing and ranting and grumbling themselves toward acts of horrid violence – such people will be known, long before they get a chance to explode in destructive spasms. We as a culture would find it harder to ignore such people, to shunt them aside and pretend the problems aren't there. So sousveillance could be a wonderful thing in terms of increasing our safety.

Ben

Neighborhood watch on steroids. Right.

But on the other hand, if some of the little old ladies on my block could see some of the things that occur in my house, they might not like it.

David

Would the nosy interventions of well-meaning do-gooders infringe on the rights of those who don't *want* to be cured? Good question. Nanny-state uniformity could be enforced by a Big Brother state, or by millions of narrow-minded Little Brothers. But it won't be, by broadminded fellow citizens. I would hope that personal eccentricity will continue to be valued, as it has increasingly been valued in the last few generations

Still, when it becomes clear that someone might be planning to do something awful, eyes will be drawn... And bad things will be deterred.

Ben

Yes, that much is clear... And the broader consequences are more fuzzy and complex, because it's a matter of how our psychology and culture would self-organize in response. There will be many distinct reactions and cases, dynamically linking together – it's going to be interesting!

David

Yes – for instance another problem is that of shy people. They simply do not *want* the glare. We would need to develop sliding scales... Much as today there is a sliding scale of expectation of privacy. People who seek fame are assumed to have less legal recourse against paparazzi or gossip articles than quiet citizens, who retain their right to sue for privacy invasion. So you can see that the principle is already being applied. In the future, folks who register high on shyness quotient will not get to prevent all sousveillance. But a burden of proof will fall on those who deliberately bug such people. The law won't enforce this. The rest of us will, by reproving the staring-bullies. It will be social, as in the tribes of old.

Ben

That's one possibility. Another possibility is that shyness, as we know it, basically becomes obsolete – and the genetic factors that lead to shyness are manifested in different ways, and/or get selected (or engineered) out of the gene pool.

David

Essentially, this is the greatest of all human experiments. In theory, sousveillance should eventually equilibrate into a situation where people (for their own sakes and because they believe in the Golden Rule, and because they will be caught if they violate it) eagerly and fiercely zoom in upon areas where others might be conniving or scheming or cheating or pursuing grossly-harmful deluded paths while *looking away* when none of these dangers apply. A socially sanctioned discretion based on "none of my business" and leaving each other alone... Because you'll want that other person to be your ally next time, when YOU are the one saying "make that guy leave me alone!"

That is where it should wind up. If we're capable of calm, or rationality and acting in our own self-interest. It is stylishly cynical for most people to guffaw, at this point and assume this is a fairy tale. I can just hear some readers muttering "Humans aren't like that!"

Well, maybe not. But I have seen plenty of evidence that we are now *more* like that than our ancestors ever imagined they could be. The goal may not be attainable. But we've already taken strides in that direction.

Ben

Hmmm... I definitely see this "best of both worlds" scenario as one possible attractor that a sousveillant society could fall into, but not necessarily the only one. I suppose we could also have convergence to other, very different attractors, for instance ones in which there really is no privacy because endless spying has become the culture; and ones in which uneasy middle-grounds between surveillance and sousveillance arise, with companies and other organizations enforcing cultures of mutual overwhelming sousveillance among their employees or members.

Just as the current set of technologies has led to a variety of different cultural "attractors" in different places, based on complex reasons.

David

This is essentially my point. The old attractor states are immensely powerful. Remember that 99% of post agricultural societies had no freedom because the oligarchs wanted it that way and they controlled the information flows. That kind of feudal-aristocratic, top-down dominance always looms, ready to take over. In fact, I think so-called Culture War is essentially an effort to discredit the "smartypants" intellectual elites who might challenge authoritarian/oligarchic attractor states, in favor of others that are based upon calm reason.

The odds have always been against the Enlightenment methodology – the core technique underlying our markets, democracy and science – called Reciprocal Accountability. On the other hand, sousveillance is nothing more or less than the final reification of that methodology. Look, I want sousveillance primarily because it will end forever the threat of top-down tyranny. But the core question you are zeroing in on, here, is a very smart one – could the cure be worse than the disease? I really don't know. In The Transparent Society I mostly pose hard questions. But the *possibility* that universal vision might lead to us all *choosing* to behave better? Well, why should it not be possible, in theory, to take it all the way, and then use it in ways that stymie its own excesses?

Ben

OK, so let's suppose this "multiple attractors" theory is right – and the "rational, calm, best of both worlds" scenario is just one possibility among many, regarding the societal consequences of sousveillance. Then an important question becomes: What can we do to nudge society toward attractors embodying the more benevolent sort of sousveillant society you envision?

David

I really don't know. I at times despair of getting traction on this, at a time when such big picture problems are obscured by everything political and sociological having to be crammed into terms of a stupid, lobotomizing so-called "left-right axis." That horribly misleading (and French) metaphor has done untold harm. If he were alive today, Adam Smith would be an enemy of Glen Beck. If you can figure out why, then maybe you can think two dimensionally.

I guess I'll regain my optimism, that people can live in a future that requires endless negotiation... When I start to see a little more negotiating going on, in present day reality.

Ben

So you're saying that IF progressive forces could escape their entrapping metaphors and align behind the Enlightenment ideals of calm pragmatism, THEN as technology advances, the odds would be fairly high of a relatively desirable "best of both worlds" sousveillance scenario coming about. But that in reality, it's hard to assess the odds of this.

David

Right. Human history suggests that the odds are low. All you need is a state of panic, over something much bigger than 9/11, for example. The needle will tip far over to *sur*veillance... The people handing paternalistic power to a state or to elites who promise protection. Sousveillance – especially the pragmatic, easy-going, self-restrained type I describe, would be an outgrowth of a confident citizenry. An extremely confident one. That's not impossible. We're *more* confident and skilled at such things than our ancestors. On the other hand, *Star Trek* we ain't. Not yet.

Ben

Sure– history isn't necessarily the best guide to the future, since a lot of things are changing rapidly in a richly interlinked way. Technology, society and psychology are all changing, in some respects at an accelerating pace, and plenty of historical precedents are being thrown by the wayside.

Out of all these factors, my mind keeps coming back to issues of psychology. The societal attractors that groups of people fall into, are obviously closely coupled with the psychological attractors that individual people fall into. If people start thinking differently, then they may come to understand the societal situation differently.... To pursue your example, if the psychology of progressive people were somehow tweaked (say, by the advent of sousveillance technology), then they might view themselves and their opposition differently. Maybe being able to see more and more of how their opponents operate, via early-stage sousveillance technology, would give progressives a better understanding of what they're up against.

I mean, it seems to me sousveillance could have a big impact on human psychology, in various ways. For instance, it seems people may lose much of their sense of shame in a sousveillant society (due to being able to see that so many others also have the same "shameful" aspects). You see this already in the younger generation – sexting implies a fairly low level of shame about body image; and the publishing of formerly-private-type details on various social media implies a fairly low level of shame about one's personal life.

David

Well, there's shame and then there's shame. It's been shown that when the public is increasingly exposed to a bizarre but harmless type of person, that type grows more-tolerated, but the exception is when that group exhibits *willful harmfulness*. Tolerance of – say – KKK-types goes *down* with every exposure. Likewise, if extensive, deliberate voyeurism is viewed as harmful... and voyeurs are always caught in the act, well, you have that "restaurant effect" we spoke of, earlier. People refrain from staring because it is shameful to be caught staring without good reason. And thus people get some privacy, even in an all-open world.

It depends on what's "shameful." And we can still sway the direction that goes. There still is time.

Ben

And the whole phenomenon of "social role-playing" (cf. Goffman's "Presentation of Self in Everyday Life" and so forth) will take on a different flavor when you know others have more capability to see through your masks – so that while social pretense will surely still exist, it may take on more of the role of explicit play-acting.

Overall, my guess is that sousveillance has a strong potential to lead to much more open, honest, direct social interactions; which would likely lead to many (surely not all!) people being more open, honest and direct *with themselves*, and thus becoming quite different than people typically are in modern societies.

David

Humans are essentially self-deluders. The mirror held up by other people helps us to perceive our own errors... Though it hurts. In his poem "To a Louse," Robert Burns said:

> *"Owad some Power the giftie gie us To see oursels as others see us! It wad frae monie a blunder free us, An' foolish notion…"*

("Oh would some power, the gift give us, to see ourselves as other see us. It would from many a blunder free us, and foolish notions…"). Or, my own aphorism is CITOKATE: Criticism Is The Only Known Antidote to Error. Too bad it tastes so awful.

Ben

I'm reminded of a book I read recently, "Stepping out of Self-Deception: The Buddha's Liberating Teaching of No-Self" (by Rodney Smith). In the Buddhist view, the self itself is a delusion, constructed by the mind for its own practical purposes and then falsely taken as a solid reality; and the antidotes to such errors and delusions are things like mindfulness and meditation and wise thinking... Which of course constitute a path with its own difficulties.

And compassion – the Buddhists place a lot of emphasis on compassion. Being linked to others at a fundamental level via compassion helps unwind the delusions of the self. And it does seem sousveillance may help here. It's easy to imagine some people would become more compassionate if they could more easily and directly observe the consequences of their actions for others.

David

This part is simply proved. Television did that, long before media went two way. There are many unknowns, but souveillance would drive compassion.

Ben

And these positive psychological changes, occurring in some, could trigger an enhanced ability to sense some of the negative or deceptive psychological characteristics of others.

David

I see no conflict between your perspective and mine. The thing that makes people reticent, paranoid, prejudiced, or eager to surrender authority to oligarchs always boils down to fear. A sousveillant society *should* equilibrate upon one with a low fear level, as we see kids display on social networks.

You say open/honest/direct. I would add "forgiving." Kids count on being able to shrug off today's stupid photos and statements, in the future. They *know* many of the things they are posting, at age 17, are dopey and even sick. They are vesting themselves in a wager that you and I will turn out to be right, and that we can build a society that chills-out about non-harmful mistakes.

Ben

Hmmm... So, as an SF author, how would you make the personalities of the characters different if they were situation in a society with powerful sousveillance technology in use?

David

Well, one key point is that *satiated* people tend to have calmer, less fearful reactions. But it also depends if they were raised in a society that teaches *satiability*. Satiation and satiability are not the same thing. Without the latter, the former is futile.

Ben

I see, and we in modern America tend to teach our kids to grow up insatiable!!

As Mignon McLaughlin put it, "Youth is not enough. And love is not enough. And success is not enough. And, if we could achieve it, enough would not be enough."

But what you seem to be saying is: if our society shifted to teach satiability, then we would be satiated (since we certainly have the technological capability to satiate ourselves in various ways!), and we'd be less fearful, and hence more likely to approach sousveillance with Enlightenment values rather than fear-driven values. It's certainly interesting to dissect the psychodynamics of the situation in that way.

David

Well, yes. I suppose. I do think we've already walked some paces down the road, from where our ancestors were. Note that when I say "satiable" I do not mean an end to ambition. I just think sane people, when they *get* what they say they wanted, should actually become happier and need it just a bit less. Then they move on to other desires.

Ben

That's certainly true in a sense, but that sort of "sanity" is fairly rare in the modern world, as you note. And it's been rare for a long time, maybe since the emergence of civilization and the end of the Stone Age cultures. Harking back to Buddhism again, this seems to relate to the "cycle of dependent origination". Desire leads to grasping leads to birth leads to death leads to confusion leads to action leads to sensation leads to desire, etc. The Buddhists complained about the ongoing cyclic phenomenon wherein the more you get the more you want and so forth – and this was thousands of years ago, before TV advertising and processed junk food and all that. But I've often wondered if this whole cycle (lack of satiability, you'd call it) is largely a transitional phenomenon – it wasn't so prominent in Stone Age societies (or so it seems from study of current Stone Age societies like the Piraha in Brazil and the Bayaka in Africa, etc.), and maybe it won't be so prominent in more advanced technological societies either. And maybe sousveillance could be part of the change by which we move back to a state of satiability.

Certainly, one might think that sousveillance could help increase satiability, because it could help people to see that others aren't really any better off than they are. For instance, my envy of the rich drastically decreased once I got to know some rich people moderately well, and saw that their lives really aren't dramatically qualitatively better than everyone else's in terms of quality of experience, even though the material surround is more impressive.

Well, I'm probably speculating too far here. I guess these particulars are hard to dissect without sinking into total wild conjecture. But anyway, there does seem to be some hope that the psychological and social factors may all come together to yield a future in which society adapts to sousveillance in a positive, rational, healthy way. No guarantee, but some hope.

David
Well, I always have hope. Most cultures burned crackpot gadflies like me at the stake for saying deliberately provocative things.

Here and now? People pay me for it. I love this civilization. I want it to survive.

Ben

Hear, hear! Our civilization is wonderful in spite of its dark aspects, and far from stable but rapidly growing and changing. While I admire the apparent satiability of Stone Age culture, I admit I don't have much desire to go back to that. I do overall tend to view our civilization as a transient phenomenon, and then the problem is how to nudge things so the next phase that it grows into is something more positive – maybe getting back some of the past things we've lost as well as getting new things we can't now imagine. You could view our civilization as a clever but unruly child, and then the hope is that it grows up into something even more interesting!

J. Storrs Hall:
Intelligent Nano Factories and Fogs

Interview first published January 2011

Computer scientist and futurist thinker J. Storrs Hall has been one of the leading lights of nanotech for some time now. Via developing concepts like utility fog[63] and weather machines, he has expanded our understanding of what nanotech may enable. Furthermore, together with nanotech icon Eric Drexler he pioneered the field of nano-CAD (nano-scale Computer Aided Design), during his service as founding Chief Scientist of Nanorex[64].

As Hall describes it, "Nanotechnology is based on the concept of tiny, self-replicating robots. The Utility Fog is a very simple extension of the idea: Suppose, instead of building the object you want atom by atom, the tiny robots linked their arms together to form a solid mass in the shape of the object you wanted? Then, when you got tired of that avant-garde coffee table, the robots could simply shift around a little and you'd have an elegant Queen Anne piece instead."

Hall – "Josh" to his friends – has also done a great deal to publicize nanotech concepts. He founded the sci.nanotech Usenet newsgroup and moderated it for ten years; wrote the book Nanofuture: What's Next for Nanotechnology[65] and led the nanotech-focused Foresight Institute from 2009-2010.

---

[63] http://autogeny.org/Ufog.html
[64] http://www.nanoengineer-1.com/content/
[65] http://www.amazon.com/Nanofuture-Nanotechnology-J-Storrs-Hall/dp/1591022878

I've known Josh mainly via his other big research interest, artificial general intelligence (AGI). His book *Beyond AI* [66] describes some of his ideas about how to create advanced AI, and also explores the history of the field and the ethical issues related to future scenarios where advanced AIs become pervasive and ultimately more intelligent than humans.

For this interview I decided to focus largely on nanotech – I was curious to pick Josh's brain a bit regarding the current state of nanotech and the future outlook. I found his views interesting and I think you will too! And toward the end of the interview I couldn't help diverging into AGI a bit as well – specifically into the potential intersection of AI and nanotech, now and in the future.

While at the present time Josh's work on AI and nanotech are fairly separate, he foresees a major convergence during the next decades. Once nanotechnology advances to the point where we have actual nanofactories, able to produce things such as flexibly configurable utility fogs, then we will need advanced AI systems to create detailed plans for the nanofactories, and to guide the utility fogs in their self-organizing rearrangements.

Ben
Nanotechnology started out with some pretty grand visions (though, I think, realistic ones) – visions of general-purpose manufacturing and computation at the molecular scale, for example. But on your website you carefully contrast "real nanotechnology, i.e. molecular machines (as opposed to films and powders re-branded "nanotech" as a buzzword)." How would you contrast the original vision of nanotech as outlined by Richard Feynman and Eric Drexler, with the relatively flourishing field of nanotech as it exists today?

---

[66] http://www.amazon.com/Beyond-AI-Creating-Conscience-Machine/dp/1591025117/

Josh

Feynman's vision of nanotech[67] was top-down and mechanical. With the somewhat arbitrary definition of nanotech as used these days, progress toward a Feynman-style nanotech won't be called "nanotechnology" until it actually gets there. So you have to look at progress in additive manufacturing, ultra-precision machining, and so forth. That's actually moving fairly well these days.

(See scientific.net[68] for a sampling of recent technical work on ultra-precision machining, or Google "ultra-precision machining" for a list of companies offering services in this area.)

Drexler's original approach to nanotech[69] was biologically based. Most people don't realize that because they associate him with his descriptions of the end products, which are of course mechanical. There's been some fairly spectacular progress in this area too, with DNA origami[70] and synthetic biology[71] and so forth.

I expect the two approaches to meet in the middle sometime in the 2020s.

Ben

I see, and when the two approaches meet, then we will definitively have "real nanotechnology" in the sense you mean on your website. Got it. Though currently the biology approach and the ultra-precision machining approach seem quite different in their underlying particulars. It will be interesting to see the extent to which these two technological approaches really do merge together – and I agree with you that it's likely to happen in some form.

---

[67] http://www.zyvex.com/nanotech/feynman.html
[68] http://www.scientific.net/AMR.69-70
[69] http://e-drexler.com/p/06/00/EOC_Cover.html
[70] http://en.wikipedia.org/wiki/DNA_origami
[71] http://en.wikipedia.org/wiki/Synthetic_biology

Next question is: In terms of the present day, what practical nanotech achievements so far impress and excite you the most?

Josh

Not much in terms of stuff you can buy, although virtually every chip in your computer is "nanotech" by the definition of the NNI (the National Nanotech Initiative[72]), so if you go by that you'd have to include the entire electronics industry.

Ben

OK, not much in terms of stuff you can buy – but what progress do you think has been made, in the last decade or so, toward the construction of "real nanotechnology" like molecular assemblers and utility fog? What recent technology developments seem to have moved us closer to this capability?

Josh

Well, in the research labs you have some really exciting work going on in DNA origami, manipulation/patterning of graphene[73], and single-atom deposition and manipulation. in the top-down direction. "Feynman's path" – you have actuators with sub-angstrom resolution and some pretty amazing results with additive e-beam sintering[74].

Ben

What about utility fog? Are we any closer now to being able to create utility fog, than we were 10 years ago? What recent technology developments seem to have moved us closer to this capability?

---

[72] http://www.nano.gov/
[73] http://arxiv.org/abs/0806.0716
[74]
   http://en.wikipedia.org/wiki/Additive_manufacturing

Josh

There've actually been some research projects in what's often called "swarm robotics" at places like CMU, although one of the key elements is to design little robots simple and cheap enough to build piles of without breaking the bank. I think we're close to being able to build golf-ball-sized Foglets – meaning full-functioned ones – if anyone wants to double the national debt. You'd have to call it "Utility Hail", I suppose.

Ben

OK, I see. So taking a Feynman-path perspective, you'd say that right now we're close to having the capability to create utility hail – i.e. swarms of golf-ball sized flying robots that interact in a coordinated way. Nobody has built it yet, but that's more a matter of cost and priorities than raw technological capability. And then it's a matter of incremental engineering improvements to make the hail-lets smaller and smaller until they become true fog-lets.

Whereas the Drexler-path approach to utility fog would be more to build upwards from molecular-scale biological interactions, somehow making more easily programmable molecules that would serve as foglets – but from that path, while there's been a lot of interesting developments, there's been less that is directly evocative of utility fog. So far.

Josh

Right. But things are developing fast and nobody can foresee the precise direction.

Ben

Now let's turn to some concrete nanotech work you played a part in. You did some great work a few years ago with NanoRex, pioneering Computer Aided Design for nanotech. What ultimately happened with that? What's the current status of computer aided design for nanotechnology? What are the main challenges in making CAD for nanotech work really effectively?

Josh

The software we built at NanoRex – NanoEngineer-1 – is open source and can be got from Sourceforge[75] if anyone wants to play with it.

But I think it's really too early for nano-CAD software to come into its prime, since the ability to design is still so far ahead of the ability to build. So software like NanoEngineer-1 where you could design and simulate gadgets from Nanosystems has no serious user base. Yet. And I'd say the same is true of other nano-CAD software that has sprung up recently.

One exception is the software that allows you to design DNA origami and similar wet approaches. But most of this is research software since the techniques are changing so fast.

There will definitely be a future for advanced nano-CAD software, but any major push will have to wait for the ability to build to catch up with the ability to design.

Ben

Speaking of things that may be "too early," I wonder if you have any thoughts on femtotechnology? Is the nanoscale as small as we can go engineering-wise, or may it be possible to create yet smaller technology via putting together nuclear particles into novel forms of matter (that are stable in everyday situations, without requiring massive gravity or temperature)?

Josh

I don't have any particular thoughts on femtotech to share. The thing about nanotech and AI are that we have natural models, molecular biology and human intelligence that show us that the goal is possible. We don't have any such thing for femtotech.

---

[75] http://sourceforge.net/projects/nanoengineer-1/

Ben

OK, fair enough... Hugo DeGaris seduced me into thinking about femtotech a bit lately, and I've come to agree with him that it may be viable – but you're right that we have no good examples of it, and in fact a lot of the relevant physics isn't firmly known yet. Definitely it makes sense for far more resources to be focused on nanotech, which is almost certainly known to be possible, so it's "just" down to engineering difficulties. Which we seem to be making good progress on!

So, switching gears yet again... You've done a lot of work on AGI as well as on nanotech – as you know my main interaction with you has been in your role as AGI researcher. How do these two threads of your work interact? Or are they two separate interests? Are there important common ideas and themes spanning your nanotech and AGI work?

Josh

There's probably some commonality at the level of a general theory of self-organizing, self-extending systems, but I'm not sure there's so much practical overlap in the near term of developing either one in the next decade. Even in robotics the apparent overlap is illusory: The kind of sensory and cognition-driven robots that are likely to be helpful in working out AGI are quite distinct from the blind pick-and-place systems that will be the first several generations of nanotech automation, I'm afraid.

Ben

And what's your view on the potential of AGI to help nanotech along? Do you think AGI will be necessary in order to make advanced nanotech like utility fog or molecular assemblers?

Josh

Not to build them but to use them to anywhere near their full potential. With either utility fog or nanofactories (and also with biotech and ordinary software) you have access to a design space that totally dwarfs the ability of humans to use it. It's easy to envision a nanofactory garage: each time you open the door you find a new car optimized for the people riding, the trip you're taking, the current price of fuel, and the weather. But who designs it; who designs the new car each time? You need an AI to do that, and one with a fairly high level of general intelligence.

Ben

Yes, I agree of course. But what kind of AGI or narrow AI do you think would be most useful for helping nanotech in this way – for designing plans to be used by nanofactories? Do you think AI based closely on the human brain would be helpful, or will one require a different sort of AI specifically designed for the kinds of reasoning involved in nanotech design? If we make an AI with sensors and actuators at the nano-scale, will its cognitive architecture need to be different than the human cognitive architecture (which is specialized somewhat for macro-level sensors and actuators)? Or can the nano-design-focused AGI have basically the same cognitive architecture as a human mind?

Josh

I think the human brain, while clearly bearing the marks of its evolutionary origin, is a remarkably general architecture. And the sensors in the human body are more like what nanotech could do than what current sensor technology can do. You have a much higher bandwidth picture of the world than any current robot; and I think that's a key element of the development of the human mind.

Ben

I definitely agree that the human brain gets a higher-bandwidth picture of the world than any current robot. And yet, one can imagine future robots with nano-scale sensors that get a much higher bandwitdh picture of the world than humans do. I can see that many parts of the human brain architecture wouldn't need to change to deal with nano-scale sensors and actuators – hierarchical perception still makes sense, as does the overall cognitive architecture of the human mind involving different kinds of memory and learning. But still, I wonder if making a mind that deals with quantum-scale phenomena effectively might require some fundamental changes from how the human mind works. But I suppose this also depends on exactly how the nanofactories work. Maybe one could create nanofactories that could be manipulated using largely classical-physics-ish reasoning, or else one could build others that would be best operated by a mind somehow specifically adapted to perception and action in the quantum world.

But as you said about applying nano-CAD, it's somewhat hard to explore these issues in detail until we have the capability to build more stuff at the nano scale. And fortunately that capability is coming along fairly rapidly!

So let's talk a bit about the timing of future technology developments. In 2001 you stated[76] that you thought the first molecular assemblers would be built between 2010 and 2020 Do you still hold to that prediction?

Josh

I'd say closer to 2020, but I wouldn't be surprised if by then there were something that could arguably be called an assembler (and is sure so to be called in the press!). On the other hand, I wouldn't be too surprised if it took another 5 or ten years beyond that, pushing it closer to 2030. We lost several years in the development of molecular nanotech [77] due to political shenanigans in the early 20-aughts and are playing catch-up to any estimates from that era.

Ben

In that same 2001 interview you also stated "I expect AI somewhere in the neighborhood of 2010.", with the term AI referring to "truly cognizant, sentient machines". It's 2011 and it

---

[76] http://www.crnano.org/interview.hall.htm

[77] http://www.kurzweilai.net/the-drexler-smalley-debate-on-molecular-assembly

seems we're not there yet. What's your current estimate, and why do you think your prior prediction didn't eventuate?

Josh

I made that particular prediction in the context of the Turing Test and expectations for AI from the 50s and 70s. Did you notice that one of the Loebner Prize[78] chatbots actually fooled the judge into thinking it was the human in the 2010 contest? We're really getting close to programs that, while nowhere near human-level general intelligence, are closing in on the level that Turing would have defended as "this machine can be said to think". IMHO. Besides chatbots, we have self-driving cars, humanoid walking robots, usable if not really good machine translation, some quite amazing machine learning and data mining technology, and literally thousands of narrow-AI applications. Pretty much anyone from the 50s would have said, yes, you have artificial intelligence now. In my book *Beyond AI* I argue that there will be at least a decade while AIs climb through the range of human intelligence. My current best guess is that that decade will be the 20s. We'll have competent robot chauffeurs and janitors before 2020, but no robot Einsteins or Shakespeares until after 2030.

Ben

Yes, I see. Of course "AI" is well known to be a moving target. As the cliché says, once something can be done, it's miraculously not considered AI anymore. We have a lot of amazing AI today already; most people don't realize how pervasive it is across various industries, from finance to military to biomedicine etc. etc. I don't really consider chatbots as any indicator of progress toward artificial general intelligence, but I do think the totality of narrow AI progress means something. I guess with both agree that it's this general progress in AI-related algorithms, together with advances in hardware and cognitive science, that's pushing us toward human-level general intelligence.

---

[78] http://www.loebner.net/Prizef/loebner-prize.html

Although the question of whether robot janitors will come before robot Einsteins is an interesting one. As you'll recall, at the AGI-09 conference we did a survey[79] of participants on the timeline to human-level AGI – asking questions about how long it will be till we have it, but also about what kinds will come first. A lot of participants roughly agreed with your time estimate, and thought we'd have human-level AGI within the next few decades. But opinions were divided about whether the janitors or the Einsteins will come first – that is, about whether the path to human-level AGI will proceed via first achieving human-like robotic body control and then going to abstract cognition; or whether we'll get advanced AGI cognition first, and humanlike robotics only afterwards. It seems you're firmly in the "robotics first camp", right?

Josh

Yes, and I explain why in my book.

I expect the progress toward true AGI to follow, to some extent, the evolutionary development of the brain, which was first and foremost a body controller whose components got copied and repurposed for general cognition. In the 70s robotics was much harder than, say, calculus, because the computers then didn't have the horsepower to handle full sensory streams but math could be squeezed into a very clean – and tiny – representation and manipulated. Nowadays we do have the horsepower to process sensory streams, and use internal representations of similar complexity. A modern GPGPU can compare two pictures in the same time an IBM 7090 took to compare two S-expressions. So, robotics is getting easier almost by the day.

But the hard part of what smart humans do is finding those representations, not using them once they're programmed in. Those AI programs weren't Newtons – they didn't invent calculus, they just used it in a very idiot-savant fashion. Feynman

---

[79] http://sethbaum.com/ac/fc_AI-Experts.html

put it this way, "But the real *glory* of science is that we can find a way of thinking such that the law is *evident*."

We already have well-understood representations for the physical world, and we can just give these to our robots. What's more, we have a good idea of what a good janitor should do, so we can quickly see where our robot is falling short, and easily evaluate the new techniques we invent to repair its deficiencies. So I'd expect rapid progress there, and indeed that's what we see. Once we have the new, working, techniques for things like recognizing coffeemakers, we'll be able to adapt them to things like recognizing promising customers, slacking employees, and so forth that form the meat of average human intelligence.

Perhaps it will be a while later before some robot muses, without being told or even asked, that the quality of mercy is not strained, but that it droppeth as the gentle rain from heaven, upon the place beneath.

Ben

Hey. Yes, I understand your perspective. As you know I don't quite agree 100%, even though I think robotics is one among several very promising paths toward human-level AGI. But I don't want to derail this interview into a debate on the degree of criticality of embodiment to AGI.

So let me sort of change the subject instead! What are you mostly working on these days? AGI? Robotics? Nanotech? Something else?

Josh

Back to AGI/robotics, after a detour into general futurism and running Foresight. Same basic approach as I described in my book *Beyond AI* and my paper[80] at AGI-09 (see also the online

---

[80] http://agi-conf.org/2009/papers/paper_22.pdf

video[81] of the talk), with an associative-memory based learning variant of Society of Mind[82]

Oh, and on the side, I'm trying to write a science fiction novel that would illustrate some of my machine ethics theories. And I should also mention that I have a couple of chapters in the new book Machine Ethics[83] coming out from Cambridge later this year.

Ben

Lots of great stuff indeed! I'm particularly curious to see how the next steps of your AGI research work out. As you say, while the nanotech field is advancing step by step along both the Feynman and Drexler type paths, once we do have that nanomanufacturing capability, we're going to need some pretty advanced AIs to run the nano-CAD software and figure out exactly what to build. Maybe I can interview you again next year and we can focus on the progress of your AGI work.

Josh

I'm looking forward to it!

---

[81] http://vimeo.com/7297453

[82] http://www.amazon.com/Society-Mind-Marvin-Minsky/dp/0671657135

[83] http://www.cambridge.org/aus/catalogue/catalogue.asp?isbn=9780521112352

Mohamad Tarifi: AGI and the Emerging Peer-to-Peer Economy

Interview first published June 2011

I first encountered Mohamad Tarifi on an AI email list, where he was discussing his deep and fascinating work on hierarchical temporal memory architectures. Using a combination of rigorous mathematics and computational experiments, he treats these constructs very broadly as an approach to both vision processing and artificial general intelligence. This is an area I've been delving into a fair bit lately due to my work hybridizing Itamar Arel's DeSTIN vision system with the OpenCog proto-AGI architecture, and I found Mohamad's work helpful in understanding this area of AI better.

But through our online conversations, it quickly became apparent that Mohamad's depth of thinking about technology and the future extends far beyond technical AI topics – for example he has developed a rich knowledge and insight regarding some themes raised in my recent interview with Linas Vepstas, e.g. the relation between open source software, AI and the rapidly transforming economy. And when Mohamad and I decided to do an interview, we chose to focus on these themes – what is happening to the world economy now, what's going to happen in the next few decades, and how do AGI and open source fit into the picture?

Ben
Before we get started with the meat of the interview, could you fill me (and the readership) in on some of the details of your background?

Mohamad
Sure. I received a bachelor of engineering in computer and communications engineering, with minors in mathematics and

physics, from the American University of Beirut, and spent 2 semesters in exchange at the University of Waterloo. After that, I did my masters on computer science at the University of Florida (UF) and helped start a research group on quantum computing with researchers from the UF and Caltech. During graduate school, I worked full-time for over 3 years doing analysis, software development, and management for several companies in the financial, video games and internet industries. I technically lead, mentored, and consulted for several young startups, industry, and research projects. Through my industry and graduate experience, I have been in involved in projects spanning many diverse areas such as bioinformatics, psychology, quantum computing, nanotechnology, economics, computational neuroscience, biomedical devices, finance, computer graphics, marketing, video game development, machine learning, complexity, and various mathematical disciplines such as combinatorics, linear algebra, statistics, and theoretical CS. I'm currently focusing full-time on the final stages of my PhD work, with the thesis titled "Foundations Towards a Computational Theory of Intelligence".

Ben

Wow, certainly a rich and diverse set of interests and experiences! But it seems that, at least in certain sectors of society, one hears that sort of story more and more. As the economy and the technology world get more complex, conventionally single-focused careers become decreasingly common.

To get the ball rolling, why don't you tell me a bit about your thinking on open source and the future of the economy... And how you see AI fitting into the picture going forward?

Mohamad

As well all know, our society is being transformed by technology. Traditional economic forces push towards optimizing labor efficiency. The automation of simple labor by machines was one such leap in efficiency that started the industrial age. Further push towards labor efficiency dictated the next technological leap to the information age. The next step in the evolution of technology is Artificial General Intelligence (AGI). The realization of the significant impact of Artificial General Intelligence (AGI) on the future of humanity motivated me to investigate the theoretical foundations for AGI as my PhD topic.

Ben

So, just to be sure it's clear to everyone – when you say AGI, you mean the whole shebang, right? Human-level AGI and then eventually general intelligence massively greater than the human level?

Mohamad

If by human level of intelligence, you mean current human level of intelligence, then I certainly think that AGI will easily surpass us. The more interesting question is what parts of our substrate is limiting us and how we will augment ourselves with AGI.

Ben

OK, great. So let's probe into the potential transformative impact of AGI a little. You isolate AGI as the next big step in human and technological evolution, and I tend to agree with you. But others might point to other technologies as critical, e.g. human brain enhancement or strong nanotech. Why do you isolate AGI in particular? Do you think human-level AGI is going to come before radical advancements in these other areas? Or do you think these other areas on their own wouldn't have such dramatic transformative potential, without AGI to accelerate them?

Mohamad

I agree with both lines of reasoning you offered. Those technologies will be synergetic, and AGI seems to be the closest feasible target.

Ben

I suppose most of our readers won't be familiar with your specific take on AI research ... so maybe it would be worthwhile for you to summarize briefly the key ideas of your investigations.

Mohamad

With the guidance of my advisers Dr. Meera Sitharam and Dr. Jeffery Ho, I'm pursuing a mathematically rigorous integrated theory intelligence from the perspective of computer science that builds upon ideas from neuroscience, machine learning, and many areas of mathematics and theoretical computer science such as approximation theory and computational learning theory. We put a preprint of the very first steps towards our integrated approach online at arxiv.org [84]. The paper introduces a fundamental circuit element for bottom-up processing that generalizes and formalizes existing hierarchical/deep models in the literature. Of course, the model will be extended and formally analyzed, along with new algorithms and experiments, in future publications.

Ben

Great, thanks. Of course that interests me a lot, and maybe we'll do another interview digging into that sometime. But for now let's focus on the economic side of things. What do you think is happening to the world economy, and how do you see AGI fitting into that?

Mohamad

The limitations of the assumptions of traditional economic theory are increasingly manifest in our society through economic inequalities, over-competitiveness, selfishness, short term thinking, and instability. Traditional currencies suffer from inherent structural instabilities, with cycles of inflation followed by crises, a pattern of samsara. To cope with the divergence between theory and reality, economists are attempting to isolate and formalize economically significant human behavioral

---

[84] http://arxiv.org/abs/1106.0357

patterns into their models. For instance, Ernst Fehr's research incorporates ideas from neuroscience to model agent behavior in various economic games. What is needed, however, is an alternative infrastructure.

Ben

A "pattern of samsara" – that's a choice phrase!

Mohamad

As you know It is as a metaphor for cycles of death and rebirth. I can go on about how this relates to the dynamics of Jungian archetypes in the collective subconscious of society, but this interview will probably never end. Plus these types of metaphors are best left a bit ambiguous since that generalizes their scope of applicability by allowing multiple interpretations. This is why for instance the ancient Vedic scripts are intentionally cryptic. You see how this discussion can quickly become too interesting to end? So I'll stop here.

Ben

Heh – but I find that, unlike you, I can't resist the urge to explicate the obvious at this point!

Presumably you mean to imply that AGI has the potential get us beyond this sort of cycle into a realm dominated by the pattern of nirvana? :-)

Mohamad

Ahahaha, yes!

Ben

Well, it's tempting to riff more about AGI and nirvana, but I suppose I'll take your cue and steer back toward our agreed thematic of the future of economy! So could you say a little more about the practical specifics of his work.

Mohamad

This is a link[85] to his research summary and some of his articles. According to his webpage, his research "focuses on the proximate pattern and the evolutionary origins of human altruism, and the interplay between social preferences, social norm and strategic interactions".

For instance, he analyzes the impact of inequity aversion in bilateral ultimatum games. In that setting, a proposer has to split a predetermined amount of money with a responder. If the responder accepts the percentage share offered by the proposer, both parties retain their share, otherwise both loose. It turns out that human responders will generally not accept the proposer's offer if it deviates far from equality, and correspondingly proposers generally offer reasonable cuts. The traditional utility function of material self-interest does not explain this behavior. By adding the assumption of inequity aversion to the utility function, he shows that the Nash equilibrium can be shifted to account for the observed human trials.

Ben

That's certainly interesting; I look forward to digging into it further.

Actually, I find the very concept of money gets a bit slippery these days, what with the complexity of the international banking system, and the emergence of attention and reputation as often more economically important than material goods, and so forth.

Mohamad

Yes, money is changing, and will likely change a lot more as AGI advances. Money is traditionally perceived as a medium for the exchange of human effort. But with the increased automation of labor, as we approach Artificial General Intelligence (AGI), the cost of effort is decreasing for the majority of the population. As the automation process continues it's exponential increase, the

---

[85] http://www.affective-sciences.org/user/66

rationale behind the concept of money is breaking down. Unless we do something about it, and I'm confident that we will, instability concerns will become even more pressing as we move towards AGI. With the breakdown of our traditional monetary system, we need a major shift in our collective consciousness as a species, to redefine the concept and meaning of work, and center it around human values.

Ben

Hmmm ... How would you specifically tie automation and its impact on the nature of money, to the recent financial crisis at the end of 2008?

Mohamad

Automation creates a more efficient and connected system. Results in complexity theory show that in network follow systems (such as ecosystems, electric circuits, or monetary transactions), there is a single measure based on connectivity that trades off efficiency with sustainability/stability. There is beautiful analogy with a concept in statistics/machine learning called the Bias-Variance trade-off: if we build a statistical model with high complexity, we are at risk of over-fitting the training data. This means that we can explain the current training data ever more precisely (efficiency), at the price of incorrectly generalizing to future test data (stability).

Ben

I guess this gets a bit technical for most of our readers, but still I'm curious: What's the measure of complexity you refer to?

Mohamad

An overview paper is "Quantifying Complexity: Resilience, efficiency, and the return of information theory", which can be found here[86]

---

[86]
http://www.lietaer.com/images/Ecological_Complexity_Final.pdf

Ben

I'm reminded of how the great social philosopher Hannah Arendt distinguished work (purposeful activity which need not be tied to material compensation, and may be done for its intrinsic or social reward) from labor (effort expended with the primary goal of gaining resources).How does her distinction resonate with your thinking?

My own thought is that AGI and allied technologies may eventually eliminate labor in her sense, but work in her sense will remain; and may shift largely toward "artistic" work, since practical work will be done by robots, and science, engineering and math will lose part of their appeal when AGI scientists can do them better (they may still be fun to do, but will lose the practicality that is now part of their appeal).

Mohamad

That resonates well with my thinking. Personally, I enjoy science and engineering. In my little spare time, I read current research papers and work on conceptual problems in theoretical physics (cosmological models, quantum field theories, quantum gravity, etc.) and many areas of mathematics (pure and applied to the various sciences). I also enjoy solving problems, programming, and hacking art such as building video games, experimenting with collaborative narrative, and real-time art. With a bit more free time on my hands, I would continue pursuing these intellectual ventures because they appeal to me aesthetically.

I think in a way this has already happened. Most people in the western world do not need to exert much physical labor. We no longer need to walk miles to a clean water source or hunt. Yet, we enjoy regular exercise, working out at the gym, and running marathons!

Ben

So money is changing meaning, and at the same time our relationship to money is changing, as we construct and enter into more complex networks of value.

Mohamad

This transition is occurring through P2P network economies afforded by the internet infrastructure, with the emergence of alternative currency systems, such as Bernard Lietaer's research[87] showing how complimentary currencies can address instability. By accounting and scaling previously informal economies, these currency systems promote a more cooperative and ethical society that tailors to our humanity. Many of these ideas have appeared in different forms throughout human history. The difference right now is that we can implement and scale them with computers and the internet.

Recently one of my friends wrote a really nice paper on P2P economies[88] along these lines, along with proofs of liquidity.

Ben

Yes, I see. This network of ideas is rapidly advancing and we're understanding more all the time, both practically and theoretically.

Could you say a little about how you see the mid-term future of alternative currency systems? Do you really think they'll become a practical alternative for a large variety of economic transactions soon? What segment of the economy do you think they'll seriously invade first?

Mohamad

Alternative currency systems are already being used on small and medium scales in niche markets, local communities and businesses (please see bellow, and Lietaer's site for some references).

What is needed is to raise global awareness in order to mainstream them and better integrate them to our legal and

---

[87] http://www.lietaer.com/
[88] http://arxiv.org/abs/1007.0515

economic infrastructure. For instance, in some cases, voting to allow taxes to be collected in alternative currencies.

By adding a new class of abundant connections to the economic network flow, specifically those derived from P2P and social interactions, these complimentary currencies increase the stability of the entire system. For example, the relative economic stability of Switzerland can be partially attributed to WIR independent complimentary currency system employed by medium and small sized businesses. The volume of transactions in the WIR increases in periods of global periods of economic stress, providing the liquidity required to keep the businesses running.

By accounting and scaling previously informal economies, these currency systems promote a more cooperative and ethical society that tailors to our humanity.

Ben

That's an intriguing claim. How exactly would these alternative currencies promote more ethical interactions?

Mohamad

I prefer to keep value judgments on the moral issues to a minimum, in order to avoid people's predetermined views on morality from interfering with the natural process of assimilating this research. On the other hand, the research results in themselves shed clear light into morally significant conundrums.

For instance, alternative P2P currencies promote long term thinking through demurrage. Positive interest rate amounts to preferring short term gains since returns on long term projects are exponentially discounted to present value. Demurrage currencies have the opposite effect, it's much more profitable to spend them on fruitful long term projects.

The Fureai Kippu is a complimentary currency employed in Japan to account for hours of service to elderly people. These credits accumulate and the user may then later use them for oneself or transfer them to someone else. This nurtures the community ethic that a trained professional can hardly match. Surveys show that elderly people consistently prefer services provided by people paid in Fureai Kippu over professionals paid in Yen.

Ben

You've also talked before about automated barter systems as part of the mix in a future economy.

Mohamad

Yes. For instance, if we choose to follow a mixed economical system: scarcity for resources which are continuously distributed in an egalitarian way, and abundance for everything else, then Barter systems for exchanging resources can be algorithmically automated.

Ben

Hmmm, but what are the advantages of barter over currency-based systems? I don't really get that. Barter just seems intrinsically inefficient compared to money. It may require multiple trades to do the same thing as can be done with a single transaction using money. Granted, using computers and the Net, multiple trades may be done rapidly. But, what's the point of using barter instead of some kind of money?

Mohamad

This was just one example of a candidate model for a P2P economy. The government can continually distribute rights to newly discovered scarce natural resources equally to its people. These resources can then be traded among the community with a Barter system. This does not mean that we trade the actual resources themselves, but the rights to them, in a P2P manner. Algorithms can then be implemented to find the most efficient series of exchanges than can maximize a collective measure of utility.

The P2P Foundation explores several of interesting alternative models that either compliments or replaces traditional social, economic, and political infrastructure. These include Ripple pay for P2P money, P2P governance, P2P manufacturing, and so on. While these models are not yet fully mainstream, they are already being implemented on large scales. The open source community is a great example of P2P production that significantly affects our daily lives.

Ben

Ah yes, I see. And I suppose open governance and open source government tie into all this as well.

Mohamad

Yes indeed! But I've already talked too much today, so perhaps we can leave this for another discussion sometime in the future?

Ben

That's fine. But there's one more thing I have to bring up before I let you go.

All this discussion about the future of the economy seems to assume a fairly optimistic future scenario where Hugo DeGaris is wrong and the artilects don't simply squash us all and co-opt our molecules for their own purposes! What is your reaction to that sort of darker, more violent future vision?

Mohamad

I think Hugo DeGaris's thinking on the topic is a bit immature. While I agree there is a threat of unfriendly AGI, I think we will respond to this threat with corresponding security precautions. These concerns affect the lives of real humans, and no less than the fate of our species. The gravity of these issues will be addressed with appropriate scrutiny and depth, rather than merely wild fantasies.

I personally like the word "human" better than "transhuman". The technological revolution is interesting precisely because it will allow us to tap into our vast potential and fully explore our humanity.

Ben

Well, I agree that Hugo's proclamations on these topics can be a bit oversimplistic sometimes (though knowing him personally, I'm aware his thinking is subtler than his proclamations sometimes make it seem). However, nevertheless, there seems some validity to the fear that an AGI with general intelligence 10x or 100x that of a human [and yeah, I know there's no accepted practical measure of general intelligence, but I'm speaking qualitatively here], might find little use for humans, and treat us in a manner similarly to how we treat, say, gnats.

Mohamad

I do agree unfriendly AGI is a concern, especially if the AGI is designed to be actively hostile.

There will be various protection mechanisms designed along with the system. For instance, we can directly monitor the progress and thinking patterns of an AGI in real time, or preemptively by performing simulations in controlled environments. In my work, I favor embedding AGI in a virtual reality world. This has many advantages including better control, duplicability, and better security.

The core premise of my view here is a bit more subtle, though, than previous distinctions made by AGI enthusiasts. To begin with, I think that reinforcement learning is not a good model for action. Action is better thought of as Active Inference[89]. With active inference, an AGI is motivated do nothing besides minimize a bound on the entropy of its sensorium. That is, without a "body", goal oriented behavior quickly breaks down. The key to sustainable behavior is to infuse some of the "soul" of humanity with AGI. I'm speaking in metaphorical terms, of course. There are many interesting concepts that we can explore

---

[89] http://inference.pdf/

here, but I think the details of this argument deserves an independent treatment, perhaps in a future interview.

Ben

Hmmm.... I don't see any *guarantee* that massively superhuman AGIs would treat us badly or dismissively, but I don't see how one can be confident of "security measures" against such a possibility. Unless you mean security measures to stop people or human-level AIs from developing to a significantly super-human level of intelligence; some kind of panopticon-powered nanny mechanism...

Mohamad

Ahahaha, I think people hopefully know by now that such protectionism does not work. It is simply too unstable a scenario.

Ben

I'm not so sure of that myself – but let's move on. I want to go back to your statement that you like the word "human" better than "transhuman". What exactly do you mean by that?

I mean, AGI technology may allow us to create "virtual humans", but it may also allow us to create powerful intelligences that are quite different from humans, don't you think? And if these are also dramatically more generally intelligent than humans, perhaps they would deserve the label "transhuman". And if humans somehow hybridize their brains with such non-human AGIs, doesn't this make them "transhuman" rather than merely "humans exploring their humanity"? At what point is a cyborg no longer a human and something else instead?

Mohamad

I agree with you on all the points above.

I think the sort of advanced beyond the current human level that you describe is possible, and in many different ways. Simple arguments can show that this can be done, trivially, as in one example: just build a human level AGI but with the capacity to replicate itself, you can then do many tasks in parallel that the attention bandwidth of a single human cannot do. As someone who has managed working up to 70 hour weeks in the industry, while attending graduate school, and keeping a relatively healthy and social lifestyle, I know how much easier things would be if I could simply fork copies of myself!

I think we're just comfortable with different words to describe the same process. To me, "Transhuman" feels detached from human concerns. I think of the future changes as a self-actualization of our human destiny. This process is continuous. I can imagine if the two of us were to have this conversation back in prehistoric times, you might have preferred to call humanity now as "Transhuman". This touches on the ancient and ever-so-interesting debate between Monism, Dualism, and Dualistic-Monism outlooks in spiritual philosophies but this is another topic that will stretch this interview.

Ben

Yes, we've covered a lot of ground already, and I thank you for taking the time to share your views with me and the readers of H+ Magazine. I look forward to following your work and your thinking as they develop!

Michael Anissimov:
The Risks of Artificial Superintelligence

Interview first published April 2011

"Existential risk" refers to the risk that the human race as a whole might be annihilated. In other words: Human extinction risk, or species-level genocide. This is an important concept because, as terrible at it would be if 90% of the human race were annihilated, wiping out 100% is a whole different matter.

Existential risk is not a fully well-defined notion, because as transhumanist technologies advance, the border between human and nonhuman becomes increasingly difficult to distinguish. If humans somehow voluntarily "transcend" their humanity and become superhuman, this seems a different sort of scenario than everyone being nuked to death. However, philosophical concerns aside, there are sufficiently many clear potential avenues to human extinction to make the "existential risk" concept valuable, including: nanotech arms races, risks associated with unethical superhuman AIs, and more mundane risks involving biological or nuclear warfare. While one doesn't wish to approach the future with an attitude of fearfulness, it's also important to keep our eyes open to the very real dangers that loom.

Michael Anissimov ranks among the voices most prominent and effective in discussing the issue of existential risk, along with other issues related to the Singularity and the future of humanity and technology. Previously he was the Media Director for the Singularity Institute for AI (now called MIRI, the Machine Intelligence Research Institute) and co-organizer of the Singularity Summit, as well as a Board member of Humanity+, Michael is also a member of the Center for Responsible

Nanotechnology's Global Task Force[90]. His blog, Accelerating Future[91], is deservedly popular, featuring in-depth discussion of many important issues related to transhumanism.

The following quote summarizes some of Michael's high-level views on existential risk:

> *I cannot emphasize this enough. If an existential disaster occurs, not only will the possibilities of extreme life extension, sophisticated nanotechnology, intelligence enhancement, and space expansion never bear fruit, but everyone will be dead, never to come back. This would be awful. Because we have so much to lose, existential risk is worth worrying about even if our estimated probability of occurrence is extremely low.*
>
> *Existential risk creates a 'loafer problem' – we always expect someone else to handle it. I assert that this is a dangerous strategy and should be discarded in favor of making prevention of such risks a central focus.*

In this dialogue I aimed to probe a little deeper, getting at Michael's views on the specific nature of the risks associated with specific technologies (especially AI), and what we might do to combat them. I knew this would be an interesting interview, because I'd talked informally with Michael about these ideas a few times before, so I knew we had many areas of disagreement along with broad areas of concurrence. So the interview veers from vehement agreement into some friendly debate – I hope you'll enjoy it!

Ben

What do you think is a reasonable short-list of the biggest existential risks facing humanity during the next century?

---

[90] http://www.crnano.org/CTF.htm
[91] http://acceleratingfuture.com/michael/blog

Michael Anissimov: The Risks of Artificial Superintelligence

Michael

1. Unfriendly AI
2. Selfish uploads
3. Molecular manufacturing arms race

Ben

What do you think are the biggest misconceptions regarding existential risk, both among individuals in the futurist community broadly conceived and among the general public?

Michael

Underestimating the significance of super-intelligence. People have a delusion that humanity is some theoretically optimum plateau of intelligence (due to brainwashing from Judeo-Christian theological ideas, which also permeate so-called "secular humanism"), which is the opposite of the truth. We're actually among the stupidest possible species smart enough to launch a civilization.

Ben

One view on the future of AI and the Singularity is that there is an irreducible uncertainty attached to the creation of dramatically greater than human intelligence. That is, in this view, there probably isn't really any way to eliminate or drastically mitigate the existential risk involved in creating superhuman AGI. So, building superhuman AI is essentially plunging into the Great Unknown and swallowing the risk because of the potential reward (where the reward may be future human benefit, or something else like the creation of aesthetically or morally pleasing superhuman beings, etc.).

Another view is that if we engineer and/or educate our AGI systems correctly, we can drastically mitigate the existential risk associated with superhuman AGI, and create a superhuman AGI that's highly unlikely to pose an existential risk to humanity. What are your thoughts on these two views? Do you have an intuition on which one is more nearly correct? (Or do you think both are wrong?) By what evidence or lines of thought is your intuition on this informed/inspired?

Michael

Would you rather your AI be based on Hitler or Gandhi?

Ben

Can I vote for a random member of Devo instead?

Michael

My point is, if you have any preference, that proves you understand that there's some correlation between a seed AI[92] and the Singularity it grows into.

Imagine that AGI were impossible. Imagine we would have to choose a human being to become the first superintelligence. Say that we knew that that human would acquire power that put her above *all others*. Say, she had the guaranteed ability to charm and brainwash everyone she came into contact with, and direct them to follow her commands. If that had to be the case, then I would advise that we choose someone with as much innate kindness and cleverness as possible. Someone that really cared for humanity as a whole, and had an appreciation for abstract philosophical and moral issues. Someone that was mostly selfless, and understood that moral realism is false[93]. Someone who followed the axioms of probability theory[94] in their reasoning. Someone who systematically makes accurate probability estimates, rather than demonstrating overconfidence, underconfidence, or framing biases.

---

[92] http://en.wikipedia.org/wiki/Seed_AI

[93] http://www.wjh.harvard.edu/~jgreene/GreeneWJH/Greene-Dissertation.pdf

[94] http://yudkowsky.net/rational/bayes

This is the future of the entire light cone we're talking about; The Galaxy, the Virgo Cluster, and beyond. Whether we like it or not, many think it's likely that the first superintelligence would become a singleton[95], implicitly taking responsibility for the future development of civilization from that point on. Now, it may be that we feel emotional aversion to the idea of a singleton, but this doesn't alter the texture of the fitness landscape. The first superintelligence, may, in fact, be able to elevate itself to singleton status quite quickly. (Say, through rapid self-replication or perhaps rapid replication of its ideas.) If it can, then we have to do our best to plan for that eventuality, whether or not we personally like it.

Ben

By the way, I wonder how you define a "singleton"?

I'm personally not sure the concept even applies to radically non-human AI systems. The individual-versus-group dichotomy works for us minds that are uniquely associated with specific physical bodies with narrow inter-body communication bandwidth, but will it hold up for AGIs?

Michael

Nick Bostrom covered this in "What is a Singleton?[96]":

> In set theory, a singleton is a set with only one member, but as I introduced the notion, the term refers to a world order in which there is a single decision-making agency at the highest level. Among its powers would be (1) the ability to prevent any threats (internal or external) to its own existence and supremacy, and (2) the ability to exert effective control over major features of its domain (including taxation and territorial allocation)

---

[95] http://www.nickbostrom.com/fut/singleton.html
[96] http://www.nickbostrom.com/fut/singleton.html

> *A democratic world republic could be a kind of singleton, as could a world dictatorship. A friendly superintelligent machine could be another kind of singleton, assuming it was powerful enough that no other entity could threaten its existence or thwart its plans. A "transcending upload" that achieves world domination would be another example.*

The idea is around a single decision-making agency. That agency could be made up of trillions of sub-agents, as long as they demonstrated harmony on making the highest level decisions, and prevented Tragedies of the Commons. Thus, a democratic world republic could be a singleton.

Ben

Well the precise definition of "harmony" in this context isn't terribly clear to me either. But at a high level, sure, I understand a singleton is supposed to have a higher degree of unity associated with its internal decision-making processes, compared to a non-singleton intelligent entity.

I think there are a lot of possibilities for the future of AGI, but I can see that a singleton AGI mind is one relatively likely outcome. So we do need to plan with this possibility in mind.

Michael

Yes, and it's conservative to assume that artificial intelligence will ascend in power very quickly, for reasons of prudence. Pursuit of the Singularity should be connected with an abundance of caution. General intelligence is the most powerful force in the universe, after all.

Human morality and "common sense" are extremely complex and peculiar information structures. If we want to ensure continuity between our world and a world with AGI, we need to transfer over our "metamorals" at high fidelity. Read the first chapter of Steven Pinker's How the Mind Works to see what I'm getting at. As Marvin Minsky said, "Easy things are hard!" "Facts" that are "obvious" to infants would be extremely complicated to specify in code. "Obvious" morality, like "don't kill people if you don't have to" is extremely complicated, but seems deceptively simple to us, because we have the brainware to compute it intuitively. We have to give AGIs goal systems that are compatible with our continued existence, or we will be destroyed [97]. Certain basic drives [98] common across many different kinds of AGIs may prove inconvenient to ourselves when the AGIs implementing them are extremely powerful and do not obey human commands.

To quote [99] the founder of the World Transhumanist Association, Nick Bostrom:

The option to defer many decisions to the superintelligence does not mean that we can afford to be complacent in how we construct the superintelligence. On the contrary, the setting up of initial conditions, and in particular the selection of a top-level goal for the superintelligence, is of the utmost importance. Our entire future may hinge on how we solve these problems.

Words worth taking seriously. We only have one chance to get this right.

Ben

This quote seems to imply a certain class of approaches to creating "superintelligence" – i.e. one in which the concept of a "top level goal" has a meaning. On the other hand, one could argue that humans don't really have top-level goals, though one can apply "top level goals" as a crude conceptual model of some aspects of what humans do. Do you think humans have top-level goals? Do you think it's necessary for a superintelligence to have a structured goal hierarchy with a top-level goal, in order for it to

---

[97] http://wiki.lesswrong.com/wiki/Paperclip_maximizer
[98] http://selfawaresystems.com/2007/11/30/paper-on-the-basic-ai-drives/
[99] http://www.nickbostrom.com/ethics/ai.html

have a reasonably high odds of turning out positively according to human standards? [By the way, my own AI architecture does involve an explicit top-level goal, so I don't ask this from a position of being radically opposed to the notion.]

Giving an AGI a moral and goal system implicitly, via interacting with it and teaching it in various particular cases, and asking it to extrapolate, would be one way to try to transmit the complex information structure of human morality and aesthetics to an AGI system without mucking with top-level goals. What do you think of this possibility?

Michael

Humans don't have hierarchical, cleanly causal goal systems (where the desirability of subgoals derives directly from their probabilistic contribution to fulfilling a supergoal). Human goals are more like a network of strange attractors, centered around sex, status, food, and comfort.

It's desirable to have an AI with a clearly defined goal system at the top because 1) I suspect that strange attractor networks converge to hierarchical goal systems in self-modifying systems, even if the network at the top is extremely complex , 2) such a goal system would be more amenable to mathematical analysis and easier to audit.

A hierarchical goal system could produce a human-like attractor network to guide its actions, if it judged that to be the best way to achieve them, but an attractor network is doomed to an imprecise approach until it crystallizes a supergoal. It's nice to have the *option* of a systematic approach to pursuing utility, rather than being necessarily limited to an unsystematic approach. I'm concerned about the introduction of randomness because random changes to complex structures tend to break those structures. For instance, if you took out a random component of a car engine and replaced it with a random machine, the car would very likely stop functioning.

My concern with putting the emphasis on teaching rather than a clear hierarchical goal system to analyze human wishes is the risk of overfitting. Most important human abilities are qualities that we are either born with or not, like the ability to do higher mathematics. Teaching, while important, seems to be more of an end-stage tweaking and icing on the cake than the meat of human accomplishment.

Ben

Hmm, OK. As I understand it, you're saying that human ethical intuition, "human ethics," are more a matter of how a human's brain is physically constructed, rather than a matter of human culture that is taught to people? And that, if you build an AGI without the correct "brain/mind structure", then no matter how well you teach it, it might not learn what you think you're teaching it?

Michael

Of course, relative to other humans, because we all have similar genetics, training seems to matter a lot. But in the scheme of all animals, our unique abilities are mostly predetermined during development of the embryo. There's a temptation to over-focus on teaching rather than creating deep goal structure because humans are dependent on teaching one another. If we had direct access to our own brains, however, the emphasis would shift very much to determining the exact structure during development in the womb, rather than teaching after most of the neural connections are already in place.

To put this another way: a person born as a psychopath will never become benevolent, no matter the training. A person born highly benevolent would have to be very intensely abused to become evil. In both cases, the inherent neurological dispositions are more of a relevant factor than the training.

Ben

One approach that's been suggested, in order to mitigate existential risks, is to create a sort of highly intelligent "AGI

Nanny" or "Singularity Steward". This would be a roughly human-level AGI system without the capability for dramatic self-modification, and with strong surveillance powers that is given the task of watching everything that humans do and trying to ensure that nothing extraordinarily dangerous happens.

One could envision this as a quasi-permanent situation, or else as a temporary fix to be put into place while more research is done regarding how to launch a Singularity safely.

What are your views on this AI Nanny scenario? Plausible or not? Desirable or not? Supposing the technology for this turns out to be feasible, what are the specific risks involved?

Michael

I'd rather not endure such a scenario. First, the name of the scenario is too prone to creating biases in appraisal of the idea. Who wants a "nanny"? Some people would evaluate the desirability of such a scenario merely based on the connotations of a nanny in all-human society, which is stupid. We're talking about qualitatively new kind of agent here, not something we can easily understand.

My main issue with the idea of an "AI Nanny" is that it would need to be practically Friendly AI-complete anyway. That is, it would have to have such a profound understanding of, and respect for, human motivations, that you'd be 99% of the way to the "goal" with such an AI anyway. Why not go all the way, and create a solution satisfactory to all, including those who are paranoid about AI nannies?

Since specifying the exact content of such a Nanny AI would be extremely difficult, it seems likely that whatever extrapolation process that could create such an AI would be suitable for building a truly Friendly AI as well. The current thinking on Friendly AI is not to create an AI that sticks around forever, but merely a stepping stone to a process that embodies humanity's wishes[100]. The AI is just an "initial dynamic" that sticks around long enough to determine the coherence between humanity's goals and implements them.

The idea is to create an AI that you actually trust. Giving control over the world to a Nanny AI would be a mistake, because you might never be able to get rid of it. I'd rather have an AI that is designed to get rid of itself once its job is done. Creating superintelligence is extremely dangerous, something you only want to do once. Get it right the first time.

I'm not sure how plausible the scenario is, it would depend upon the talents of the programmer. I'm concerned that it would be possible. I think it's very likely that if we take stupid shortcuts, we'll regret it. Some classes of AI might be able to keep us from dying indefinitely, under conditions we find boring or otherwise suboptimal. Imagine a civilization frozen with today's people and technology forever. I enjoy the present world, but I can imagine it might get boring after a few thousand years.

Ben

Hmmm. You say "an AI Nanny would need to be practically Friendly AI-complete anyway." Could you justify that assertion a little more fully? That's not so clear to me.

It seems that understanding and respecting human motivations is one problem, whereas maintaining one's understanding and goal system under radical self-modification, using feasible computational resources, is another problem. I'm not sure why you think solution of the first problem implies being near to the solution of the second problem.

Michael

"AI Nanny" implies an AI that broadly respects humans in ways that do not lead to our death or discomfort, but yet restricts our freedom in some way. My point is that if it's already gone so far

---

[100] http://singularity.org/files/CEV.pdf

to please us, why not go the full length and give us our freedom? Is it really that impossible to please humans, even if you have more computing power and creativity at your disposal than thousands of human races?

The solution of the first problem implies being near to the second problem because large amounts of self-modification and adjustment would be necessary for an artificial intelligence to respect human desires and needs to begin with. Any AI sophisticated enough to do so well will already have engaged in more mental self-modifications than any human being could dream of. Prepping an AI for open-ended self-improvement after that will be an additional challenging task.

I'm worried that if we created an AI Nanny, we wouldn't be able to get rid of it. So, why not create a truly Friendly AI instead, one that we can trust, that provides us with long-term happiness and satisfaction as a benevolent partner to the human race? Pretty simple.

If we had a really benevolent human and an uploading machine, would we ask them to just kickstart the Singularity, or have them be a Nanny first? I would presume the former, so why would we ask an AI to be a nanny? If we trust the AI like a human, it can do everything a human can do, and it's the best available entity to do this, so why not let it go ahead and enhance its own intelligence in an open-ended fashion? If we can trust a human then we can trust an intelligently built friendly AGI even more.

I suspect that by the time we have an AI smart enough to be a nanny, it would be able to build itself MNT computers the size of the Hoover Dam, and solve the problem of post-Nanny AI.

Ben

The disconnect between my question and your answer seems to be that I think a Nanny AI (without motivation for radical self-modification) might be much easier to make than a superintelligence which keeps its goals stable under radical self-modification (and has motivation for radical self-motivation). Yeah, if you think the two problems are roughly of equal difficulty, I see why you'd see little appeal in the Nanny AI scenario.

Michael

Yes, there's the disagreement. I'd be interested in hearing your further arguments for why one is so much harder than the other, or why the AI couldn't make the upgrade to itself with little human help at that point.

Ben

Why do I think a Nanny AI is easier than a superintelligent radically self-modifying AGI? All a Nanny AI needs to do is to learn to distinguish desirable from undesirable human situations (which is probably a manageable supervised classification problem), and then deal with a bunch of sensors and actuators distributed around the world in an intelligent way. A super-AI on the other hand has got to deal with situations much further from those foreseeable or comprehensible by its creators, which poses a much harder design problem IMO...

Michael

Again, overfitting. Perhaps it's desirable to me to risk my life walking a tightrope, an intelligently designed Nanny AI would be forced to stop me. The number of special cases is too extreme, it requires real understanding. Otherwise, why bother with a Nanny AI, why not create an AI that just fulfills the wishes of a single steward human? I'd rather have someone with real understanding in control than a stupid AI that is very powerful but lacks basic common sense and the ability to change or listen. If something is more powerful than me, I want it to be more philosophically sophisticated and benevolent than me, or I'm likely against it. (Many people seem to be against the idea of any agents more powerful than them, period, which is ironic because I don't exactly see them trying to improve their power much either.)

Ben

Yeah, it requires real understanding – but IMO much *less* real understanding than maintaining goal stability under radical self-modification...

As to why to prefer a Nanny AI to a human dictator, it's because for humans power tends to corrupt, whereas for AGIs it won't necessarily. And democratic human institutions appear probably unable to handle the increasingly dangerous technologies that are going to emerge in the next N decades...

Another point is that personally, as an AI researcher, I feel fairly confident I know how to make a nanny AI that could work. I also know how to make an AGI that keeps its goal system stable under radical self-modification – BUT the catch is, this design (GOLEM) uses infeasibly much computational resources. I do NOT know how to make an AGI that keeps its goal system stable under radical self-modification and runs using feasible computational resources, and I wonder if designing such a thing is waaaay beyond our current scientific / engineering / mathematical capability.

Michael

How on Earth could you be confident that you could create a Nanny AI with your current knowledge? What mechanism would you design to allow us to break the AI's control once we were ready? Who would have control of said mechanism?

Ben

The AI would need to break its control itself, once it was convinced we had a plausible solution to the problem of building a self-modifying AGI with a stable goal system. Human dictators don't like to break their own control (though it's happened) but AIs needn't have human motivations...

Michael

If the AI can make this judgment, couldn't it build a solution itself? An AI with the power to be a Nanny would have more cognitive resources than all human beings that have ever lived.

Ben

Very often in computer science and ordinary human life, *recognizing* a solution to a problem is much easier than actually finding the solution... This is the basis of the theory of NP-completeness for example... And of course the scientists who validated Einstein's General Relativity theory mostly would not have been able to originate it themselves...

However, one quite likely outcome IMO is that the Nanny AI, rather than human scientists, is the one to find the solution to "radically self-modifying AGI with a stable goal system"... Thus obsoleting itself ;)... Or maybe it will be a collaboration of the Nanny with the global brain of internetworked brain-enhanced humans... It should be fun to find out which path is ultimately taken!

Michael

OK, that last clause sounds vaguely similar to my SIAI colleague Eliezer Yudkowsky's idea of "Coherent Extrapolated Volition", which I'm sure you're familiar with. CEV, also, would start out gathering info and would only create the AI to replace itself once it was confident it did the extrapolation well. This would involve inferring the structure and function of human brains.

Ben

I admit the details of the CEV idea, as Eliezer explained it, never made that much sense to me. But maybe you're interpreting it a little differently.

Michael

CEV just means extrapolating what people want. We do this every day. Even salamander mothers do this for their children. The cognitive machinery that takes sense perceptions of an agent and infers what that agent wants from those sense perceptions is the same sort of machinery that would exist in CEV.

Ben

Hmmm... That's not exactly how I understood it from Eliezer's paper...What he said is:

> *Our coherent extrapolated volition is our wish if we knew more, thought faster, were more the people we wished we were, had grown up farther together; where the extrapolation converges rather than diverges, where our wishes cohere rather than interfere; extrapolated as we wish that extrapolated, interpreted as we wish that interpreted.*

To me this seems a lot different than just "extrapolating what people want".

Michael

The thing is that if all these qualifications were not here, extrapolation would lead to suboptimal outcomes. For instance, you must have made decisions for your children that were more in alignment with what they would want if they were smarter. If you made judgments in alignment with their actual preferences (like wanting to eat candy all day – I don't know your kids but I know a lot of kids would do this), they would suffer for it in the longer term.

If extrapolations were made taking into account only our current knowledge and not our knowledge if we knew more, really bad things could happen.

If extrapolations were made based on our human-characteristic thinking speeds, rather than the long-term equilibria of thinking that we would reach immediately if we thought faster, bad things could happen.

If extrapolations were made based on the people we are often petty and under the control of short-term motivations, rather than who we wished we were, bad things could happen.

The same for each element above. I can understand why you might disagree with some of the above bullet points, but it's hard to imagine how you could disagree with the notion of volition extrapolation in general. It is a marvel of human intelligence and inference that no sometimes means yes and yes means no. An AI without a subtle extrapolation process will miss this entirely, and make choices for us that are too closely related to our current states, providing lock-in that we would never have chosen if we were superintelligences.

Salamanders extrapolate preferences. Human extrapolate preferences. Superintelligences will extrapolate preferences. Each new level of intelligence demands a higher level of wisdom for extrapolation. A superintelligence that uses human-level extrapolation algorithms to fulfill wishes would be a menace.

Ben

Hmmm... Well, to me, the idea of "what Bob would want, if Bob were more of the person Bob wishes he was" is a bit confusing, because "the person Bob wishes he was" is a difficult sort of abstraction. Bob doesn't usually genuinely know what kind of person he wishes he was. He may think he wishes he was an enlightened Zen master – and if he became an enlightened Zen master he might be quite contented that way – but yet the fact that he never took action to become that Zen master during his life in spite of many opportunities, still indicates that large parts of him didn't really want that... The notion of "the person you want to be" isn't well-defined at all...

And looking at cases where different peoples' wishes cohere is pretty dangerous too. For one thing you'd likely be throwing out your valued rationality, as that is certainly not something on which most peoples' wishes cohere. Belief in reincarnation is more likely to make it into the CEV of the human race than rationality.

And looking at the desires you would have if you were "more of the person you wish you were" is probably going to exaggerate the incoherence problem, not mitigate it. Most religious people will wish they were even MORE religious and god-fearing, so I'd be less coherent with their ideals than I am with their actual selves...

Michael

"More like the person I wish I was" is not a difficult abstraction. I have many desired modifications to my mental architecture, and I would prefer than an AI take that into account in its judgments. If Bob has dark thoughts at times, Bob wouldn't want those dark thoughts to be integrated into the preference aggregation algorithm. It seems simple enough. Without this explicit precaution, said dark thoughts that Bob would choose to be excluded from the preference aggregator would be included anyhow.

The list of items in the definition of Coherente Extrapolated Volition is a way of saying to the AI, "take this into account too". The alternative is to not take them into account. That seems bad, because these items obviously should be taken into account.

Ben

Hmmm... But I think separating out certain thoughts of Bob's from the rest of his mind is not a very easy or well-defined task either. The human mind is not a set of discretely defined logical propositions; it's a strangely tangled web of inter-definitions, right?

You may not remember, but a couple years ago, in reaction to some of my concerns with the details of the CEV idea, some time ago I defined an alternative called Coherent Aggregated

Volition (CAV)[101]. The idea was to come up with a CEV-like idea that seemed less problematic. Basically, CAV is about trying to find a goal that maximizes several criteria together, such as consistency, and matching closely on average to what a lot of people want, and compactness, and supported-ness by evidence. I guess this fits within your broad notion of "extrapolation" but it seems rather different from CEV the way Eliezer stated it.

Michael

This is extremely undesirable, because the idea is not to average out our existing preferences, but to create something new that can serve as a foundation for the future. Similarity to existing gobses should not be a criterion. We are not trying to create a buddy but a Transition Guide, a massively powerful entity whose choices will *de facto* set the stage for our entire future light cone. The tone of this work, especially w/ regards to the language about the averaging of existing preferences, does not take the AI's role as Transition Guide sufficiently into account.

Ben

Hmmm... I just think the Transition Guide should start from where we are, not from where we (or our optimization algorithm) speculate we might be if our ideal were much smarter, etc...

I think we should provide a superhuman AI initially with some basic human values, not with some weird wacky far-out extrapolation that bears no noticeable resemblance to current human values... Sure, a little extrapolation is needed, but *only* a little...

Still, I guess I can agree with you that "some idea in the vague vicinity of what you guys call CEV" is probably valuable. I could write a detailed analysis of why I think the details of Eli's CEV

---

[101] http://multiverseaccordingtoben.blogspot.com/2010/03/coherent-aggregated-volition-toward.html

paper are non-workable, but that would take a long day's work, and I don't want to put a day into it right now. Going back and forth any further on particular points in this dialogue probably isn't productive...

Perhaps one way to express the difference is that

- CAV wants to get at the core of real, current human values, as manifested in real human life
- CEV wants to get at the core of "what humans would like their values to be", as manifested in what we would like our life to be if we were all better people who were smarter and knew more

Does that feel right to you as a rough summary of the distinction?

Michael

Yes, the difference between CEV and CAV that you list makes sense.

Ben

I suppose what I'm getting at is – I think there is a depth and richness and substance to our current, real human values; whereas I think that "what we would like our values to be" is more of a web of fictions, not to be relied upon...

That is – to put it more precisely – suppose one used some sort of combination of CEV and CAV, i.e. something that took into account both current human values, and appropriately extrapolated human values. And suppose this combination was *confidence-weighted*, i.e. it paid more attention to those values that were known with more confidence. Then, my suspicion is that when the combination was done, one would find that the CAV components dominated the CEV components, because of the huge uncertainties in the latter... But it seems you have a very different intuition on this...

But anyway… I can see this is a deep point that goes beyond the scope of a dialogue like this one! I think I'd better move on to my next question!

So here goes… Another proposal that's been suggested, to mitigate the potential existential risk of human-level or superhuman AGIs, is to create a community of AGIs and have them interact with each other, comprising a society with its own policing mechanisms and social norms and so forth. The different AGIs would then keep each other in line. A "social safety net" so to speak. Steve Omohundro, for example, has been a big advocate of this approach.

What are your thoughts on this sort of direction?

Michael
Creating a community of AIs is just a way of avoiding the challenge of making an AI you trust.

Create an AI you trust, then worry about the rest. An AI that understands us. Someone we can call our friend, our ally. An agent really on our side. Then, the rest will follow. The key is not to see AI as an alien but as a potential friend. Necessarily regard AI as our enemy, and we will fail.

The universe is not fundamentally Darwinian. If the nice guy has all the weapons, all the control, then the thief and the criminal are screwed. We can defeat death. That's an affront to Darwinian evolution if there ever was one. We don't need to balance AIs off against each other. We need a proxy to a process that represents what we want.

An AI is not like a human individual. A single AI could actually be legion. A single AI might split its awareness into thousands of threads as necessary. Watson does this all the time when it searches through many thousands of documents in parallel.

We don't need to choose exactly what we want right away. We can just set up a system that leaves the option open in the future. Something that doesn't lock us into any particular local maxima in the fitness space.

Eliezer nailed this question in 2001. He really had his thumb right on it. From the FAQ section of Creating Friendly AI[102]:

- *Aren't individual differences necessary to intelligence?*
- *Isn't a society necessary to produce ideas? Isn't capitalism necessary for efficiency?*

Individual differences and the free exchange of ideas are necessary to human intelligence because it's easy for a human to get stuck on one idea and then rationalize away all opposition. One scientist has one idea, but then gets stuck on it and becomes an obstacle to the next generation of scientists. A Friendly seed AI doesn't rationalize. Rationalization of mistaken ideas is a complex functional adaptation that evolves in imperfectly deceptive social organisms. Likewise, there are limits to how much experience any one human can accumulate, and we can't share experiences with each other. There's a limit to what one human can handle, and so far it hasn't been possible to build bigger humans.

As for the efficiency of a capitalist economy, in which the efforts of self-interested individuals sum to a (sort of) harmonious whole: Human economies are constrained to be individualist because humans are individualist. Local selfishness is not the miracle that enables the marvel of a globally efficient economy; rather, all human economies are constrained to be locally selfish in order to work at all. Try to build an economy in defiance of human nature, and it won't work. This constraint is not necessarily something that carries over to minds in general.

---

[102] http://singinst.org/upload/CFAI.html#_blank

Michael Anissimov: The Risks of Artificial Superintelligence

Humans have to cooperate because we're idiots when we think alone due to egocentric biases. The same does not necessarily apply to AGI. You can make an AGI that avoids egocentric biases from the get-go. People have trouble understanding this because they are anthropomorphic and find it impossible to imagine such a being. They can doubt, but the empirical evidence will flood us from early experiments in infrahuman AI. You can call me on this in 2020.

Ben

Hmmm... I understand your and Eliezer's view, but then some other deep futurist thinkers such as Steve Omohundro feel differently. As I understand it, Steve feels that a trustable community might be easier to create than a trustable "singleton" mind. And I don't really think he is making any kind of simplistic anthropomorphic error. Rather (I think) he thinks that cooperation between minds is a particular sort of self-organizing dynamic that implicitly gives rise to certain emergent structures (like morality for instance) via its self-organizing activity...

But maybe this is just a disagreement about AGI architecture – i.e. you could say he wants to architect an AGI as a community of relatively distinct subcomponents, whereas you want to architect it with a more unified internal architecture?

Michael

Possibly! Maybe the long-term outcome will be determined by which is easier to build, and my preferences don't matter because one is just inherently more practical. Most successful AIs, like Watson and Google, seem to have unified architectures. The data-gathering infrastructure may be distributed but the decision-making, while probabilistic, is more or less unified.

Ben

One final question, then. What do you think society could be doing now to better mitigate against existential risks... From AGI or from other sources? More specific answers will be more fully appreciated ;)

Michael

Create a human-friendly superintelligence. The arguments for why this is a good idea have been laid out numerous times, and is the focus of Nick Bostrom's essay "Ethical Issues in Advanced Artificial Intelligence [103]". An increasing majority of transhumanists are adopting this view.

Ben

Hmmm... Do you have any evidence in favor of the latter sentence? My informal impression is otherwise, though I don't have any evidence about the matter... I wonder if your impression is biased due to your own role in the Singularity Institute, a community that certainly does take that view. I totally agree with your answer, of course – but I sure meet a lot of futurist who don't.

Michael

I mostly communicate with transhumanists that are not already Singularitarians because I am an employee that interfaces with the outside-of-SIAI community. I also have gotten hundreds of emails from media coverage over the past two years. If anything my view is biased towards transhumanists in the Bay Area, of which there are many, but not necessarily transhumanists with direct contact with those who already advocate friendly superintelligence.

Ben

Very interesting indeed. Well I hope you're right that this trend exists, it's good to see the futurist community adopting a more and more realistic view on these issues, which as you point out are as important as anything on the planet these days.

---

[103] http://www.nickbostrom.com/ethics/ai.html

Muehlhauser & Goertzel: Rationality, Risk, and the Future of AGI

Interview first published May 2012

Over the years, I've had a complex, often contentious relationship with the California non-profit called the "Singularity Institute for AI" (SIAI). On the one hand, SIAI helped provide seed funding for my OpenCog AGI project. On the other hand, numerous individuals associated with SIAI have gone on record stating that if I were to succeed with building an AGI according to my designs, the result would be the near-inevitable death of all human beings. My frustration with the latter line of thinking has been posted online in an essay titled "The Singularity Idea's Scary Idea, and Why I Don't Buy It," which attracted a fair bit of commentary.

SIAI was founded by futurist pundit / AGI theorist Eliezer Yudkowsky; but Eliezer is a scientist and writer rather than a manager, so SIAI has always had other business management, and has undergone a few management changes since its inception. In 2012, shortly after Luke Muehlauser came on board SIAI as its new director, he offered to conduct a structured dialogue with me on issues of central concern to him and his SIAI colleagues: the safety of future AGIs, the nature of rational thinking, and so forth. I agreed. The results are given here, and are fairly representative of the ongoing dialogue that has occurred between myself and the SIAI community over the years.

Luke
Ben, I'm glad you agreed to discuss artificial general intelligence (AGI) with me. There is much on which we agree, and much on which we disagree, so I think our dialogue will be informative to many readers, and to us!

Let us begin where we agree. We seem to agree that:

- Involuntary death is bad, and can be avoided with the right technology.
- Humans can be enhanced by merging with technology.
- Humans are on a risky course in general, because powerful technologies can destroy us, humans are often stupid, and we are unlikely to voluntarily halt technological progress.
- AGI is likely this century.
- AGI will, after a slow or hard takeoff, completely transform the world. It is a potential existential risk, but if done wisely, could be the best thing that ever happens to us.
- Careful effort will be required to ensure that AGI results in good things for humanity.

Next: Where do we disagree?

Two people might agree about the laws of thought most likely to give us an accurate model of the world, but disagree about which conclusions those laws of thought point us toward. For example, two scientists may use the same scientific method but offer two different models that seem to explain the data.

Or, two people might disagree about the laws of thought most likely to give us accurate models of the world. If that's the case, it will be no surprise that we disagree about which conclusions to draw from the data. We are not shocked when scientists and theologians end up with different models of the world.

Unfortunately, I suspect you and I disagree at the more fundamental level — about which methods of reasoning to use when seeking an accurate model of the world.

I sometimes use the term "Technical Rationality[104]" to name my methods of reasoning. Technical Rationality is drawn from two sources: (1) the laws of logic, probability theory, and decision theory [105], and (2) the cognitive science [106] of how our haphazardly evolved brains *fail to reason in accordance* with the laws of logic, probability theory, and decision theory.

Ben, at one time you tweeted[107] a William S. Burroughs quote: "Rational thought is a failed experiment and should be phased out." I don't know whether Burroughs meant by "rational thought" the specific thing *I* mean by "rational thought," or what exactly you meant to express with your tweet, but I suspect we have different views of how to reason successfully about the world.

I think I would understand your way of thinking about AGI better if I understand your way of thinking about *everything*. For example: do you have reason to reject the laws of logic, probability theory, and decision theory? Do you think we disagree about the basic findings of the cognitive science of humans? What are your positive recommendations for reasoning about the world?

Ben

Firstly, I don't agree with that Burroughs quote that "Rational thought is a failed experiment" – I mostly just tweeted it because I thought it was funny! I'm not sure Burroughs agreed with his own quote either. He also liked to say that linguistic communication was a failed experiment, introduced by women to

---

[104] http://facingthesingularity.com/2011/from-skepticism-to-technical-rationality/

[105] http://facingthesingularity.com/2011/the-laws-of-thought/

[106] http://facingthesingularity.com/2011/the-crazy-robots-rebellion/

[107] https://twitter.com/#!/bengoertzel/statuses/48066955227299840

help them oppress men into social conformity. Yet he was a writer and loved language. He enjoyed being a provocateur.

However, I do think that some people overestimate the power and scope of rational thought. That is the truth at the core of Burroughs' entertaining hyperbolic statement...

I should clarify that I'm a huge fan of logic, reason and science. Compared to the average human being, I'm practically obsessed with these things! I don't care for superstition, nor for unthinking acceptance of what one is told; and I spent a lot of time staring at data of various sorts, trying to understand the underlying reality in a rational and scientific way. So I don't want to be pigeonholed as some sort of anti-rationalist!

However, I do have serious doubts both about the power and scope of rational thought in general – and much more profoundly, about the power and scope of what you call "technical rationality."

First of all, about the limitations of rational thought broadly conceived – what one might call "semi-formal rationality", as opposed to "technical rationality." Obviously this sort of rationality has brought us amazing things, like science and mathematics and technology. Hopefully it will allow us to defeat involuntary death and increase our IQs by orders of magnitude and discover new universes, and all sorts of great stuff. However, it does seem to have its limits.

It doesn't deal well with consciousness – studying consciousness using traditional scientific and rational tools has just led to a mess of confusion. It doesn't deal well with ethics either, as the current big mess regarding bioethics indicates.

And this is more speculative, but I tend to think it doesn't deal that well with the spectrum of "anomalous phenomena" – precognition, extrasensory perception, remote viewing, and so forth. I strongly suspect these phenomena exist, and that they can be understood to a significant extent via science – but also that science as presently constituted may not be able to grasp them fully, due to issues like the mindset of the experimenter helping mold the results of the experiment.

There's the minor issue of Hume's problem of induction, as well. I.e., the issue that, in the rational and scientific world-view, that we have no rational reason to believe that any patterns observed in the past will continue into the future. This is an ASSUMPTION, plain and simple – an act of faith. Occam's Razor (which is one way of justifying and/or further specifying the belief that patterns observed in the past will continue into the future) is also an assumption and an act of faith. Science and reason rely on such acts of faith, yet provide no way to justify them. A big gap.

Furthermore -- and more to the point about AI – I think there's a limitation to the way we now model intelligence, which ties in with the limitations of the current scientific and rational approach. I have always advocated a view of intelligence as "achieving complex goals in complex environments", and many others have formulated and advocated similar views. The basic idea here is that, for a system to be intelligent it doesn't matter WHAT its goal is, so long as its goal is complex and it manages to achieve it. So the goal might be, say, reshaping every molecule in the universe into an image of Mickey Mouse. This way of thinking about intelligence, in which the goal is strictly separated from the methods for achieving it, is very useful and I'm using it to guide my own practical AGI work.

On the other hand, there's also a sense in which reshaping every molecule in the universe into an image of Mickey Mouse is a STUPID goal. It's somehow out of harmony with the Cosmos – at least that's my intuitive feeling. I'd like to interpret intelligence in some way that accounts for the intuitively apparent differential stupidity of different goals. In other words, I'd like to be able to deal more sensibly with the interaction of scientific and normative knowledge. This ties in with the incapacity of science and reason in their current forms to deal with ethics effectively, which I mentioned a moment ago.

I certainly don't have all the answers here – I'm just pointing out the complex of interconnected reasons why I think contemporary science and rationality are limited in power and scope, and are going to be replaced by something richer and better as the growth of our individual and collective minds progresses. What will this new, better thing be? I'm not sure – but I have an inkling it will involve an integration of "third person" science/rationality with some sort of systematic approach to first-person and second-person experience.

Next, about "technical rationality" – of course that's a whole other can of worms. Semi-formal rationality has a great track record; it's brought us science and math and technology, for example. So even if it has some limitations, we certainly owe it some respect! Technical rationality has no such track record, and so my semi-formal scientific and rational nature impels me to be highly skeptical of it! I have no reason to believe, at present, that focusing on technical rationality (as opposed to the many other ways to focus our attention, given our limited time and processing power) will generally make people more intelligent or better at achieving their goals. Maybe it will, in some contexts – but what those contexts are, is something we don't yet understand very well.

I provided consulting once to a project aimed at using computational neuroscience to understand the neurobiological causes of cognitive biases in people employed to analyze certain sorts of data. This is interesting to me; and it's clear to me that in this context, minimization of some of these textbook cognitive biases would help these analysts to do their jobs better. I'm not sure how big an effect the reduction of these biases would have on their effectiveness, though, relative to other changes one might make, such as changes to their workplace culture or communication style.

On a mathematical basis, the justification for positing probability theory as the "correct" way to do reasoning under uncertainty relies on arguments like Cox's axioms, or de Finetti's Dutch Book arguments. These are beautiful pieces of math, but when you talk about applying them to the real world, you run into a lot of problems regarding the inapplicability of their assumptions. For instance, Cox's axioms include an axiom specifying that (roughly speaking) multiple pathways of arriving at the same conclusion must lead to the same estimate of that conclusion's truth value. This sounds sensible but in practice it's only going to be achievable by minds with arbitrarily much computing capability at their disposal. In short, the assumptions underlying Cox's axioms, de Finetti's arguments, or any of the other arguments in favor of probability theory as the correct way of reasoning under uncertainty, do NOT apply to real-world intelligences operating under strictly bounded computational resources. They're irrelevant to reality, except as inspirations to individuals of a certain cast of mind.[108]

(An aside is that my own approach to AGI does heavily involve probability theory – using a system I invented called Probabilistic Logic Networks, which integrates probability and logic in a unique way. I like probabilistic reasoning. I just don't venerate it as uniquely powerful and important. In my OpenCog AGI architecture, it's integrated with a bunch of other AI methods, which all have their own strengths and weaknesses.)

---

[108] Ben Goertzel has undertaken some technical work aimed at overcoming these limitations, e.g. in a paper presented at the AGI-12 conference showing how to extend a certain analogue of Cox's Theorem to have weaker and more realistic assumptions ("Probability Theory Ensues from Assumptions of Approximate Consistency: A Simple Derivation and its Implications for AGI", see http://www.agroparistech.fr/mia/agiworkshop/papers/Goertzel.pdf). But this remains exploratory research work.

So anyway – there's no formal mathematical reason to think that "technical rationality" is a good approach in real-world situations; and "technical rationality" has no practical track record to speak of. And ordinary, semi-formal rationality itself seems to have some serious limitations of power and scope.

So what's my conclusion? Semi-formal rationality is fantastic and important and we should use it and develop it – but also be open to the possibility of its obsolescence as we discover broader and more incisive ways of understanding the universe (and this is probably moderately close to what William Burroughs really thought). Technical rationality is interesting and well worth exploring but we should still be pretty skeptical of its value, at this stage – certainly, anyone who has supreme confidence that technical rationality is going to help humanity achieve its goals better, is being rather IRRATIONAL :-)

In this vein, I've followed the emergence of the Less Wrong community with some amusement and interest. One ironic thing I've noticed about this community of people intensely concerned with improving their personal rationality is: by and large, these people are already hyper-developed in the area of rationality, but underdeveloped in other ways! Think about it – who is the prototypical Less Wrong meet-up participant? It's a person who's very rational already, relative to nearly all other humans – but relatively lacking in other skills like intuitively and empathically understanding other people. But instead of focusing on improving their empathy and social intuition (things they really aren't good at, relative to most humans), this person is focusing on fine-tuning their rationality more and more, via reprogramming their brains to more naturally use "technical rationality" tools! This seems a bit imbalanced. If you're already a fairly rational person but lacking in other aspects of human development, the most rational thing may be NOT to focus on honing your "rationality fu" and better internalizing Bayes' rule into your subconscious – but rather on developing those other aspects of your being... An analogy would be: If you're very physically strong but can't read well, and want to self-improve, what should you focus your time on? Weight-lifting or literacy? Even if greater strength is ultimately your main goal, one argument for focusing on literacy would be that you might read something that would eventually help you weight-lift better! Also you might avoid getting ripped off by a corrupt agent offering to help you with your bodybuilding career, due to being able to read your own legal contracts. Similarly, for people who are more developed in terms of rational inference than other aspects, the best way for them to become more rational might be for them to focus time on these other aspects (rather than on fine-tuning their rationality), because this may give them a deeper and broader perspective on rationality and what it really means.

Finally, you asked: "What are your positive recommendations for reasoning about the world?" I'm tempted to quote Nietzsche's Zarathustra, who said "Go away from me and resist Zarathustra!" I tend to follow my own path, and generally encourage others to do the same. But I guess I can say a few more definite things beyond that...

To me it's all about balance. My friend Allan Combs calls himself a "philosophical Taoist" sometimes; I like that line! Think for yourself; but also, try to genuinely listen to what others have to say. Reason incisively and analytically; but also be willing to listen to your heart, gut and intuition, even if the logical reasons for their promptings aren't apparent. Think carefully through the details of things; but don't be afraid to make wild intuitive leaps. Pay close mind to the relevant data and observe the world closely and particularly; but don't forget that empirical data is in a sense a product of the mind, and facts only have meaning in some theoretical context. Don't let your thoughts be clouded by your emotions; but don't be a feeling-less automaton, don't make judgments that are narrowly rational but fundamentally unwise. As Ben Franklin said, "Moderation in all things, including moderation."

Luke

I whole-heartedly agree that there are plenty of Less Wrongers who, rationally, should spend less time studying rationality and more time practicing social skills and generic self-improvement methods! This is part of why I've written so many scientific self-help posts for Less Wrong: Scientific Self Help[109], How to Beat Procrastination[110], How to Be Happy[111], Rational Romantic Relationships[112], and others. It's also why I taught social skills classes at our two summer 2011 rationality[113] camps[114].

Back to rationality. You talk about the "limitations" of "what one might call 'semi-formal rationality', as opposed to 'technical rationality.' "But I argued for technical rationality, so: what are the limitations of technical rationality? Does it, as you claim for "semi-formal rationality", fail to apply to consciousness or ethics or precognition? Does Bayes' Theorem remain true when looking at the evidence about awareness, but cease to be true when we look at the evidence concerning consciousness or precognition?

You talk about technical rationality's lack of a track record, but I don't know what you mean. Science was successful because it did a much better job of approximating perfect Bayesian probability theory than earlier methods did (e.g. faith, tradition), and science can be even *more* successful when it tries *harder* to

---

[109] http://lesswrong.com/lw/3nn/scientific_selfhelp_the_state_of_our_knowledge/
[110] http://lesswrong.com/lw/3w3/how_to_beat_procrastination/
[111] http://lesswrong.com/lw/4su/how_to_be_happy/
[112] http://lesswrong.com/lw/63i/rational_romantic_relationships_part_1/
[113] http://singinst.org/blog/2011/06/21/rationality-minicamp-a-success/
[114] http://lesswrong.com/lw/4wm/rationality_boot_camp/

approximate perfect Bayesian probability theory — see *The Theory That Would Not Die*[115].

You say that "minimization of some of these textbook cognitive biases would help [some] analysts to do their jobs better. I'm not sure how big an effect the reduction of these biases would have on their effectiveness, though, relative to other changes one might make, such as changes to their workplace culture or communication style." But this misunderstands what I mean by Technical Rationality[116]. If teaching these people about cognitive biases would lower the expected value of some project, then technical rationality would recommend against teaching these people cognitive biases (at least, for the purposes of maximizing the expected value of that project). Your example here is a case of Straw Man Rationality[117]. (But of course I didn't expect you to know everything I meant by Technical Rationality in advance! Though, I did provide a link to an explanation of what I meant by Technical Rationality in my first entry, above.)

The same goes for your dismissal of probability theory's foundations. You write that "In short, the assumptions underlying Cox's axioms, de Finetti's arguments, or any of the other arguments in favor of probability theory as the correct way of reasoning under uncertainty, do NOT apply to real-world intelligences operating under strictly bounded computational resources." Yes, we don't have infinite computing power. The point is that Bayesian probability theory is an *ideal* that can be *approximated* by finite beings. That's why science works better than faith — it's a better approximation of using probability theory to reason about the world, even though science is still a long way from a *perfect* use of probability theory.

---

[115] http://www.amazon.com/Theory-That-Would-Not-Die/dp/0300169698/

[116] http://facingthesingularity.com/2011/the-laws-of-thought/

[117] http://facingthesingularity.com/2011/why-spock-is-not-rational/

Re: goals. Your view of intelligence as "achieving complex goals in complex environments" does, as you say, assume that "the goal is strictly separated from the methods for achieving it." I prefer a definition of intelligence as "efficient cross-domain optimization[118]," but my view — like yours — also assumes that goals (what one values) are logically orthogonal to intelligence (one's ability to achieve what one values).

Nevertheless, you report an intuition that shaping every molecule into an image of Mickey Mouse is a "stupid" goal. But I don't know what you mean by this. A goal of shaping every molecule into an image of Mickey Mouse is an instrumentally intelligent goal if one's utility function will be maximized that way. Do you mean that it's a stupid goal according to *your* goals? But of course. This is, moreover, what we would expect your intuitive judgments to report, even if your intuitive judgments are irrelevant to the math of what would and wouldn't be an instrumentally intelligent goal for a *different* agent to have. The Mickey Mouse goal is "stupid" only by a definition of that term that is *not* the opposite of the explicit definitions either of us gave "intelligent," and it's important to keep that clear. And I certainly don't know what "out of harmony with the Cosmos" is supposed to mean.

Re: induction. I won't dive into that philosophical morass here. Suffice it to say that my views on the matter are expressed pretty well in Where Recursive Justification Hits Bottom[119], which is also a direct response to your view that science and reason are great but rely on "acts of faith."

Your final paragraph sounds like common sense, but it's too vague, as I think you would agree. One way to force a more precise answer to such questions is to think of how you'd

---

[118] http://facingthesingularity.com/2011/playing-taboo-with-intelligence/

[119] http://lesswrong.com/lw/s0/where_recursive_justification_hits_bottom/

program it into an AI. As Daniel Dennett said, "AI makes philosophy honest."

How would you program an AI to learn about reality, if you wanted it to have the most accurate model of reality possible? You'd have to be a bit more specific than "Think for yourself; but also, try to genuinely listen to what others have to say. Reason incisively and analytically; but also be willing to listen to your heart, gut and intuition..."

My own answer to the question of how I would program an AI to build as accurate a model of reality as possible is this: I would build it to use computable approximations of perfect technical rationality — that is, roughly: Computable approximations of Solomonoff induction[120] and Bayesian decision theory[121].

Ben

Bayes Theorem is "always true" in a formal sense, just like 1+1=2, obviously. However, the connection between formal mathematics and subjective experience, is not something that can be fully formalized.

Regarding consciousness, there are many questions, including what counts as "evidence." In science we typically count something as evidence if the vast majority of the scientific community counts it as a real observation – so ultimately the definition of "evidence" bottoms out in social agreement. But there's a lot that's unclear in this process of classifying an observation as evidence via a process of social agreement among multiple minds. This unclarity is mostly irrelevant to the study of trajectories of basketballs, but possibly quite relevant to study of consciousness.

Regarding psi, there are lots of questions, but one big problem is that it's possible the presence and properties of a psi effect may

---

[120] http://www.vetta.org/documents/disSol.pdf
[121] http://wiki.lesswrong.com/wiki/Decision_theory

depend on the broad context of the situation whether the effect takes place. Since we don't know which aspects of the context are influencing the psi effect, we don't know how to construct controlled experiments to measure psi. And we may not have the breadth of knowledge nor the processing power to reason about all the relevant context to a psi experiment, in a narrowly "technically rational" way... I do suspect one can gather solid data demonstrating and exploring psi (and based on my current understanding, it seems this has already been done to a significant extent by the academic parapsychology community; see a few links I've gathered[122]), but I also suspect there may be aspects that elude the traditional scientific method, but are nonetheless perfectly real aspects of the universe.

Anyway both consciousness and psi are big, deep topics, and if we dig into them in detail, this interview will become longer than either of us has time for...

About the success of science – I don't really accept your Bayesian story for why science was successful. It's naive for reasons much discussed by philosophers of science. My own take on the history and philosophy of science, from a few years back, is in an article I wrote called "Science, Probability and Human Nature: A Sociological/Computational/Probabilist Philosophy of Science "[123] (that article was the basis for a chapter in my book *The Hidden Pattern*, also). My goal in that essay was "a philosophical perspective that does justice to both the *relativism and sociological embeddedness* of science, and the *objectivity and rationality* of science." It seems you focus overly much on the latter and ignore the former. That article tries to explain why probabilist explanations of real-world science are quite partial and miss a lot of the real story. But again, a long

---

[122] http://wp.goertzel.org/?page_id=154

[123] http://www.goertzel.org/dynapsyc/2004/PhilosophyOfScience_v2.htm

debate on the history of science would take us too far off track from the main thrust of this interview.

About technical rationality, cognitive biases, etc. – I did read that blog entry that you linked, on technical rationality. Yes, it's obvious that focusing on teaching an employee to be more rational, need not always be the most rational thing for an employer do, even if that employer has a purely rationalist worldview. For instance, if I want to train an attack dog, I may do better by focusing limited time and attention on increasing his strength rather than his rationality. My point was that there's a kind of obsession with rationality in some parts of the intellectual community (e.g. some of the Less Wrong orbit) that I find a bit excessive and not always productive. But your reply impels me to distinguish two ways this excess may manifest itself:

- Excessive belief that rationality is the "right" way to solve problems and think about issues, in principle
- Excessive belief that, tactically, explicitly employing tools of technical rationality is a good way to solve problems in the real world

Psychologically I think these two excesses probably tend to go together, but they're not logically coupled. In principle, someone could hold either one, but not the other.

This sort of ties in with your comments on science and faith. You view science as progress over faith – and I agree if you interpret "faith" to mean "traditional religions." But if you interpret "faith" more broadly, I don't see a dichotomy there. Actually, I find the dichotomy between "science" and "faith" unfortunately phrased, since science itself ultimately relies on acts of faith also. The "problem of induction" can't be solved, so every scientist must base his extrapolations from past into future based on some act of faith. It's not a matter of science vs. faith, it's a matter of what one chooses to place one's faith in. I'd personally rather place faith in the idea that patterns observed in the past will likely continue into the future (as one example of a science-friendly article of faith), than in the word of some supposed "God" – but I realize I'm still making an act of faith.

This ties in with the blog post "Where Recursive Justification Hits Bottom" that you pointed out. It's pleasant reading but of course doesn't provide any kind of rational argument against my views. In brief, according to my interpretation, it articulates a faith in the process of endless questioning:

> The important thing is to **hold nothing back** in your criticisms of how to criticize; nor should you regard the unavoidability of loopy justifications as a warrant of **immunity from questioning**.

I share that faith, personally.

Regarding approximations to probabilistic reasoning under realistic conditions (of insufficient resources), the problem is that we lack rigorous knowledge about what they are. We don't have any theorems telling us what is the best way to reason about uncertain knowledge, in the case that our computational resources are extremely restricted. You seem to be assuming that the best way is to explicitly use the rules of probability theory, but my point is that there is no mathematical or scientific foundation for this belief. You are making an act of faith in the doctrine of probability theory! You are assuming, because it feels intuitively and emotionally right to you, that even if the conditions of the arguments for the correctness of probabilistic reasoning are NOT met, then it still makes sense to use probability theory to reason about the world. But so far as I can tell, you don't have a RATIONAL reason for this assumption, and certainly not a mathematical reason.

Re your response to my questioning the reduction of intelligence to goals and optimization – I understand that you are intellectually committed to the perspective of intelligence in terms of optimization or goal-achievement or something similar to that. Your response to my doubts about this perspective basically just re-asserts your faith in the correctness and completeness of this sort of perspective. Your statement:

> *The Mickey Mouse goal is "stupid" only by a definition of that term that is not the opposite of the explicit definitions either of us gave "intelligent," and it's important to keep that clear*

basically asserts that it's important to agree with your opinion on the ultimate meaning of intelligence!

On the contrary, I think it's important to explore alternatives to the understanding of intelligence in terms of optimization or goal-achievement. That is something I've been thinking about a lot lately. However, I don't have a really crisply-formulated alternative yet.

As a mathematician, I tend not to think there's a "right" definition for anything. Rather, one explains one's definitions, and then works with them and figures out their consequences. In my AI work, I've provisionally adopted a goal-achievemement based understanding of intelligence – and have found this useful, to a significant extent. But I don't think this is the true and ultimate way to understand intelligence. I think the view of intelligence in terms of goal-achievement or cross-domain optimization misses something, which future understandings of intelligence will encompass. I'll venture that in 100 years the smartest beings on Earth will have a rigorous, detailed understanding of intelligence according to which:

> *The Mickey Mouse goal is "stupid" only by a definition of that term that is not the opposite of the explicit definitions either of us gave "intelligent," and it's important to keep that clear*

seems like rubbish...

As for your professed inability to comprehend the notion of "harmony with the Cosmos" – that's unfortunate for you, but I guess trying to give you a sense for that notion, would take us way too far afield in this dialogue!

Finally, regarding your complaint that my indications regarding how to understanding the world are overly vague. Well – according to Franklin's idea of "Moderation in all things, including moderation", one should also exercise moderation in precisiation. Not everything needs to be made completely precise and unambiguous (fortunately, since that's not feasible anyway).

I don't know how I would program an AI to build as accurate a model of reality as possible, if that were my goal. I'm not sure that's the best goal for AI development, either. An accurate model in itself, doesn't do anything helpful. My best stab in the direction of how I would ideally create an AI, if computational resource restrictions were no issue, is the GOLEM design that I described in a talk at the AGI-12 conference at Oxford[124]. GOLEM is a design for a strongly self-modifying superintelligent AI system, which might plausibly have the possibility of retaining its initial goal system through successive self-modifications. However, it's unclear to me whether it will ever be feasible to build.

You mention Solomonoff induction and Bayesian decision theory. But these are abstract mathematical constructs, and it's unclear to me whether it will ever be feasible to build an AI system fundamentally founded on these ideas, and operating within feasible computational resources. Marcus Hutter and Juergen Schmidhuber and their students are making some efforts in this direction, and I admire those researchers and this body of work, but don't currently have a high estimate of its odds of leading to any sort of powerful real-world AGI system.

Most of my thinking about AGI has gone into the more practical problem of how to make a human-level AGI:

---

[124] http://goertzel.org/GOLEM.pdf

- using currently feasible computational resources
- that will most likely be helpful rather than harmful in terms of the things I value
- that will be smoothly extensible to intelligence beyond the human level as well

For this purpose, I think Solomonoff induction and probability theory are useful, but aren't all-powerful guiding principles. For instance, in the OpenCog[125] AGI design (which is my main practical AGI-oriented venture at present), there is a component doing automated program learning of small programs – and inside our program learning algorithm, we explicitly use an Occam bias, motivated by the theory of Solomonoff induction. And OpenCog also has a probabilistic reasoning engine, based on the math of Probabilistic Logic Networks (PLN). I don't tend to favor the language of "Bayesianism", but I would suppose PLN should be considered "Bayesian" since it uses probability theory (including Bayes rule) and doesn't make a lot of arbitrary, a priori distributional assumptions. The truth value formulas inside PLN are based on an extension of imprecise probability theory, which in itself is an extension of standard Bayesian methods (looking at envelopes of prior distributions, rather than assuming specific priors).

In terms of how to get an OpenCog system to model the world effectively and choose its actions appropriately, I think teaching it and working together with it, will be be just as important as programming it. Right now the project is early-stage and the OpenCog design is maybe 50% implemented. But assuming the design is right, once the implementation is done, we'll have a sort of idiot savant childlike mind, that will need to be educated in the ways of the world and humanity, and to learn about itself as well. So the general lessons of how to confront the world, that I cited above, would largely be imparted via interactive

---

[125] http://opencog.org

experiential learning, vaguely the same way that human kids learn to confront the world from their parents and teachers.

Drawing a few threads from this conversation together, it seems that

- I think technical rationality, and informal semi-rationality, are both useful tools for confronting life – but not all-powerful
- I think Solomonoff induction and probability theory are both useful tools for constructing AGI systems – but not all-powerful

Whereas you seem to ascribe a more fundamental, foundational basis to these particular tools.

Luke

To sum up, from my point of view:

- We seem to disagree on the applications of probability theory. For my part, I'll just point people to "A Technical Explanation of Technical Explanation[126]".
- I don't think we disagree much on the "sociological embeddedness" of science.
- I'm also not sure how much we really disagree about Solomonoff induction and Bayesian probability theory. I've already agreed that no machine will use these in practice because they are not computable — my point was about their provable optimality given infinite computation (subject to qualifications; see AIXI[127]).

You've definitely misunderstood me concerning "intelligence." This part is definitely not true: "I understand that you are

---

[126] http://yudkowsky.net/rational/technical
[127] http://www.amazon.com/Universal-Artificial-Intelligence-Algorithmic-Probability/dp/3642060528/

intellectually committed to the perspective of intelligence in terms of optimization or goal-achievement or something similar to that. Your response assumes the correctness and completeness of this sort of perspective." Intelligence as efficient cross-domain optimization is merely a *stipulated* definition. I'm happy to use other definitions of intelligence in conversation, so long as we're clear which definition we're using when we use the word. Or, we can replace the symbol with the substance[128] and talk about "efficient cross-domain optimization" or "achieving complex goals in complex environments" without ever using the word "intelligence."

My point about the Mickey Mouse goal was that when you called the Mickey Mouse goal "stupid," this could be confusing, because "stupid" is usually the opposite of "intelligent," but your use of "stupid" in that sentence didn't seem to be the opposite of *either* definition of intelligence we each gave. So I'm still unsure what you mean by calling the Mickey Mouse goal "stupid."

This topic provides us with a handy transition away from philosophy of science and toward AGI. Suppose there was a machine with a vastly greater-than-human capacity for either "achieving complex goals in complex environments" or for "efficient cross-domain optimization." And suppose that machine's utility function would be maximized by reshaping every molecule into a Mickey Mouse shape. We can avoid the tricky word "stupid," here. The question is: Would that machine decide to change its utility function so that it doesn't continue to reshape every molecule into a Mickey Mouse shape? I think this is unlikely, for reasons discussed in Omohundro (2008)[129].

---

[128]
http://lesswrong.com/lw/nv/replace_the_symbol_with_the_substance/

[129]
http://selfawaresystems.files.wordpress.com/2008/01/ai_drives_final.pdf

I suppose a natural topic of conversation for us would be your October 2010 blog post *The Singularity Institute's's Scary Idea (and Why I Don't Buy It)*[130]. Does that post still reflect your views pretty well, Ben?

Ben

About the hypothetical uber-intelligence that wants to tile the cosmos with molecular Mickey Mouses – I truly don't feel confident making any assertions about a real-world system with vastly greater intelligence than me. There are just too many unknowns. Sure, according to certain models of the universe and intelligence that may seem sensible to some humans, it's possible to argue that a hypothetical uber-intelligence like that would relentlessly proceed in tiling the cosmos with molecular Mickey Mouses. But so what? We don't even know that such an uber-intelligence is even a possible thing – in fact my intuition is that it's not possible.

Why may it not be possible to create a very smart AI system that is strictly obsessed with that stupid goal? Consider first that it may not be possible to create a real-world, highly intelligent system that is strictly driven by explicit goals – as opposed to being partially driven by implicit, "unconscious" (in the sense of deliberative, reflective consciousness) processes that operate in complex interaction with the world outside the system. Because pursuing explicit goals is quite computationally costly compared to many other sorts of intelligent processes. So if a real-world system is necessarily not wholly explicit-goal-driven, it may be that intelligent real-world systems will naturally drift away from certain goals and toward others. My strong intuition is that the goal of tiling the universe with molecular Mickey Mouses would fall into that category. However, I don't yet have any rigorous argument to back this up. Unfortunately my time is limited, and while I generally have more fun theorizing and philosophizing

---

[130] http://multiverseaccordingtoben.blogspot.com/2010/10/singularity-institutes-scary-idea-and.html

than working on practical projects, I think it's more important for me to push toward building AGI than just spend all my time on fun theory. (And then there's the fact that I have to spend a lot of my time on applied narrow-AI projects to pay the mortgage and put my kids through college, etc.)

But anyway – you don't have any rigorous argument to back up the idea that a system like you posit is possible in the real-world, either! And SIAI has staff who, unlike me, are paid full-time to write and philosophize... And they haven't come up with a rigorous argument in favor of the possibility of such a system, either. Although they have talked about it a lot, though usually in the context of paperclips rather than Mickey Mouses.

So, I'm not really sure how much value there is in this sort of thought-experiment about pathological AI systems that combine massively intelligent practical problem solving capability with incredibly stupid goals (goals that may not even be feasible for real-world superintelligences to adopt, due to their stupidity).

Regarding the concept of a "stupid goal" that I keep using, and that you question – I admit I'm not quite sure how to formulate rigorously the idea that tiling the universe with Mickey Mouses is a stupid goal. This is something I've been thinking about a lot recently. But here's a first rough stab in that direction: I think that if you created a highly intelligent system, allowed it to interact fairly flexibly with the universe, and also allowed it to modify its top-level goals in accordance with its experience, you'd be very unlikely to wind up with a system that had this goal (tiling the universe with Mickey Mouses). That goal is out of sync with the Cosmos, in the sense that an intelligent system that's allowed to evolve itself in close coordination with the rest of the universe, is very unlikely to arrive at that goal system. I don't claim this is a precise definition, but it should give you some indication of the direction I'm thinking in...

The tricky thing about this way of thinking about intelligence, which classifies some goals as "innately" stupider than others, is that it places intelligence not just in the system, but in the system's broad relationship to the universe – which is something that science, so far, has had a tougher time dealing with. It's unclear to me which aspects of the mind and universe science, as we now conceive it, will be able to figure out. I look forward to understanding these aspects more fully...

About my blog post on "The Singularity Institute's Scary Idea" – yes, that still reflects my basic opinion. After I wrote that blog post, Michael Anissimov – a long-time SIAI staffer and zealot whom I like and respect greatly – told me he was going to write up and show me a systematic, rigorous argument as to why "an AGI not built based on a rigorous theory of Friendliness is almost certain to kill all humans" (the proposition I called "SIAI's Scary Idea"). But he hasn't followed through on that yet – and neither has Eliezer or anyone associated with SIAI.

Just to be clear, I don't really mind that SIAI folks hold that "Scary Idea" as an intuition. But I find it rather ironic when people make a great noise about their dedication to rationality, but then also make huge grand important statements about the future of humanity, with great confidence and oomph, that are not really backed up by any rational argumentation. This ironic behavior on the part of Eliezer, Michael Anissimov and other SIAI principals doesn't really bother me, as I like and respect them and they are friendly to me, and we've simply "agreed to disagree" on these matters for the time being. But the reason I wrote that blog post is because my own blog posts about AGI were being trolled by SIAI zealots (not the principals, I hasten to note) leaving nasty comments to the effect of "SIAI has proved that if OpenCog achieves human level AGI, it will kill all humans. "Not only has SIAI not proved any such thing, they have not even made a clear rational argument!

As Eliezer has pointed out to me several times in conversation, a clear rational argument doesn't have to be mathematical. A clearly formulated argument in the manner of analytical philosophy, in favor of the Scary Idea, would certainly be very interesting. For example, philosopher David Chalmers recently wrote a carefully argued philosophy paper arguing for the plausibility of a Singularity in the next couple hundred years. It's somewhat dull reading, but it's precise and rigorous in the manner of analytical philosophy, in a manner that Kurzweil's writing (which is excellent in its own way) is not. An argument in favor of the Scary Idea, on the level of Chalmers' paper on the Singularity, would be an excellent product for SIAI to produce. Of course a mathematical argument might be even better, but that may not be feasible to work on right now, given the state of mathematics today. And of course, mathematics can't do everything – there's still the matter of connecting mathematics to everyday human experience, which analytical philosophy tries to handle, and mathematics by nature cannot.

My own suspicion, of course, is that in the process of trying to make a truly rigorous analytical philosophy style formulation of the argument for the Scary Idea, the SIAI folks will find huge holes in the argument. Or, maybe they already intuitively know the holes are there, which is why they have avoided presenting a rigorous write-up of the argument!

Luke

I'll drop the stuff about Mickey Mouse so we can move on to AGI. Readers can come to their own conclusions on that.

Your main complaint seems to be that the Singularity Institute hasn't written up a clear, formal argument (in analytic philosophy's sense, if not the mathematical sense) in defense of our major positions — something like Chalmers' " The Singularity: A Philosophical Analysis[131]" but more detailed.

I have the same complaint. I wish "The Singularity: A Philosophical Analysis" had been written 10 years ago, by Nick Bostrom and Eliezer Yudkowsky. It *could* have been written back then. Alas, we had to wait for Chalmers to speak at

---

[131] http://consc.net/papers/singularityjcs.pdf

Singularity Summit 2009 and then write a paper based on his talk. And if it wasn't for Chalmers, I fear we'd *still* be waiting for such an article to exist. (Bostrom's forthcoming *Superintelligence* book should be good, though.)

I was hired by the Singularity Institute in September 2011 and have since then co-written two papers explaining some of the basics: "Intelligence Explosion: Evidence and Import[132]" and "The Singularity and Machine Ethics[133]." I also wrote the first ever outline of categories of open research problems in AI risk, cheekily titled "So You Want to Save the World[134]." I'm developing other articles on "the basics" as quickly as I can. I would love to write more, but alas, I'm also busy being the Singularity Institute's Executive Director.

Perhaps we could reframe our discussion around the Singularity Institute's latest exposition of its basic ideas, "Intelligence Explosion: Evidence and Import"? Which claims in that paper do you most confidently disagree with, and why?

Ben

You say "Your main complaint seems to be that the Singularity Institute hasn't written up a clear, formal argument (in analytic philosophy's sense, if not the mathematical sense) in defense of our major positions ". Actually, my main complaint is that some of SIAI's core positions seem almost certainly WRONG, and yet they haven't written up a clear formal argument trying to justify these positions – so it's not possible to engage SIAI in rational discussion on their apparently wrong positions. Rather, when I try to engage SIAI folks about these wrong-looking positions (e.g. the "Scary Idea" I mentioned above), they tend to point me

---

[132] http://commonsenseatheism.com/wp-content/uploads/2012/02/Muehlhauser-Salamon-Intelligence-Explosion-Evidence-and-Import.pdf

[133] http://commonsenseatheism.com/wp-content/uploads/2011/11/Muehlhauser-Helm-The-Singularity-and-Machine-Ethics-draft.pdf

[134] http://lukeprog.com/SaveTheWorld.html

to Eliezer's blog ("Less Wrong") and tell me that if I studied it long and hard enough, I would find that the arguments in favor of SIAI's positions are implicit there, just not clearly articulated in any one place. This is a bit frustrating to me – SIAI is a fairly well-funded organization involving lots of smart people and explicitly devoted to rationality, so certainly it should have the capability to write up clear arguments for its core positions... If these arguments exist. My suspicion is that the Scary Idea, for example, is not backed up by any clear rational argument – so the reason SIAI has not put forth any clear rational argument for it, is that they don't really have one! Whereas Chalmers' paper carefully formulated something that seemed obviously true...

Regarding the paper "Intelligence Explosion: Evidence and Import", I find its contents mainly agreeable – and also somewhat unoriginal and unexciting, given the general context of 2012 Singularitarianism. The paper's three core claims that:

> *(1) there is a substantial chance we will create human-level AI before 2100, that (2) if human-level AI is created, there is a good chance vastly superhuman AI will follow via an "intelligence explosion," and that (3) an uncontrolled intelligence explosion could destroy everything we value, but a controlled intelligence explosion would benefit humanity enormously if we can achieve it*

are things that most "Singularitarians" would agree with. The paper doesn't attempt to argue for the "Scary Idea" or Coherent Extrapolated Volition or the viability of creating some sort of provably Friendly AI – or any of the other positions that are specifically characteristic of SIAI. Rather, the paper advocates what one might call "plain vanilla Singularitarianism." This may be a useful thing to do, though, since after all there are a lot of smart people out there who aren't convinced of plain vanilla Singularitarianism.

I have a couple small quibbles with the paper, though. I don't agree with Omohundro's argument about the "basic AI drives" (though Steve is a friend and I greatly respect his intelligence and deep thinking). Steve's argument for the inevitability of these drives in AIs is based on evolutionary ideas, and would seem to hold up in the case that there is a population of distinct AIs competing for resources – but the argument seems to fall apart in the case of other possibilities like an AGI mindplex (a network of minds with less individuality than current human minds, yet not necessarily wholly blurred into a single mind – rather, with reflective awareness and self-modeling at both the individual and group level).

Also, my "AI Nanny" concept is dismissed too quickly for my taste (though that doesn't surprise me!). You suggest in this paper that to make an AI Nanny, it would likely be necessary to solve the problem of making an AI's goal system persist under radical self-modification. But you don't explain the reasoning underlying this suggestion (if indeed you have any). It seems to me – as I say in my "AI Nanny" paper – that one could probably make an AI Nanny with intelligence significantly beyond the human level, without having to make an AI architecture oriented toward radical self-modification. If you think this is false, it would be nice for you to explain why, rather than simply asserting your view. And your comment "Those of us working on AI safety theory would very much appreciate the extra time to solve the problems of AI safety..." carries the hint that I (as the author of the AI Nanny idea) am NOT working on AI safety theory. Yet my GOLEM design is a concrete design for a potentially Friendly AI (admittedly not computationally feasible using current resources), and in my view constitutes greater progress toward actual FAI than any of the publications of SIAI so far. (Of course, various SIAI associated folks often allude that there are great, unpublished discoveries about FAI hidden in the SIAI vaults – a claim I somewhat doubt, but can't wholly dismiss of course....)

Anyway, those quibbles aside, my main complaint about the paper you cite is that it sticks to "plain vanilla Singularitarianism" and avoids all of the radical, controversial positions that distinguish SIAI from myself, Ray Kurzweil, Vernor Vinge and the rest of the Singularitarian world. The crux of the matter, I suppose is the third main claim of the paper,

> *(3) An uncontrolled intelligence explosion could destroy everything we value, but a controlled intelligence explosion would benefit humanity enormously if we can achieve it.*

This statement is hedged in such a way as to be almost obvious. But yet, what SIAI folks tend to tell me verbally and via email and blog comments is generally far more extreme than this bland and nearly obvious statement.

As an example, I recall when your co-author on that article, Anna Salamon, guest lectured in the class on Singularity Studies that my father and I were teaching at Rutgers University in 2010. Anna made the statement, to the students, that (I'm paraphrasing, though if you're curious you can look up the online course session which was saved online and find her exact wording) "If a superhuman AGI is created without being carefully based on an explicit Friendliness theory, it is ALMOST SURE to destroy humanity." (i.e., what I now call SIAI's Scary Idea)

I then asked her (in the online class session) why she felt that way, and if she could give any argument to back up the idea.

She gave the familiar SIAI argument that, if one picks a mind at random from "mind space", the odds that it will be Friendly to humans are effectively zero.

I made the familiar counter-argument that this is irrelevant, because nobody is advocating building a random mind. Rather, what some of us are suggesting is to build a mind with a Friendly-looking goal system, and a cognitive architecture that's roughly human-like in nature but with a non-human-like propensity to choose its actions rationally based on its goals, and then raise this AGI mind in a caring way and integrate it into society. Arguments against the Friendliness of random minds are irrelevant as critiques of this sort of suggestion.

So, then she fell back instead on the familiar (paraphrasing again) "OK, but you must admit there's a non-zero risk of such an AGI destroying humanity, so we should be very careful – when the stakes are so high, better safe than sorry!"

I had pretty much the same exact argument with SIAI advocates Tom McCabe and Michael Anissimov on different occasions; and also, years before, with Eliezer Yudkowsky and Michael Vassar – and before that, with (former SIAI Executive Director) Tyler Emerson. Over all these years, the SIAI community maintains the Scary Idea in its collective mind, and also maintains a great devotion to the idea of rationality, but yet fails to produce anything resembling a rational argument for the Scary Idea – instead repetitiously trotting out irrelevant statements about random minds!!

What I would like is for SIAI to do one of these three things, publicly:

1. Repudiate the Scary Idea
2. Present a rigorous argument that the Scary Idea is true
3. State that the Scary Idea is a commonly held intuition among the SIAI community, but admit that no rigorous rational argument exists for it at this point

Doing any one of these things would be intellectually honest. Presenting the Scary Idea as a confident conclusion, and then backing off when challenged into a platitudinous position equivalent to "there's a non-zero risk… Better safe than sorry…", is not my idea of an intellectually honest way to do things.

Why does this particular point get on my nerves? Because I don't like SIAI advocates telling people that I, personally, am on a R&D course where if I succeed I am almost certain to destroy humanity! That frustrates me. I don't want to destroy humanity; and if someone gave me a rational argument that my work was most probably going to be destructive to humanity, I would stop doing the work and do something else with my time! But the fact that some other people have a non-rational intuition that my work, if successful, would be likely to destroy the world – this doesn't give me any urge to stop. I'm OK with the fact that some other people have this intuition – but then I'd like them to make clear, when they state their views, that these views are based on intuition rather than rational argument. I will listen carefully to rational arguments that contravene my intuition – but if it comes down to my intuition versus somebody else's, in the end I'm likely to listen to my own, because I'm a fairly stubborn maverick kind of guy...

Luke

Ben, you write:

> When I try to engage SIAI folks about these wrong-looking positions (e.g. the "Scary Idea" I mentioned above), they tend to point me to Eliezer's blog ("Less Wrong") and tell me that if I studied it long and hard enough, I would find that the arguments in favor of SIAI's positions are implicit there, just not clearly articulated in any one place. This is a bit frustrating to me...

No kidding! It's very frustrating to me, too. That's one reason I'm working to clearly articulate the arguments in one place, starting with articles on the basics like "Intelligence Explosion: Evidence and Import."

I agree that "Intelligence Explosion: Evidence and Import" covers only the basics and does not argue for several positions associated uniquely with the Singularity Institute. It is, after all, the opening chapter of a book on the intelligence explosion, not the opening chapter of a book on the Singularity Institute's ideas!

I wanted to write that article first, though, so the Singularity Institute could be clear on the basics. For example, we needed to be clear that: (1) we are not Kurzweil, and our claims don't depend on his detailed storytelling or accelerating change curves, that (2) technological prediction is hard, and we are not being naively overconfident about AI timelines, and that (3) intelligence explosion is a convergent outcome of many paths the future may take. There is also much content that is not found in, for example, Chalmers' paper: (a) an overview of methods of technological prediction, (b) an overview of speed bumps and accelerators toward AI, (c) a reminder of breakthroughs like AIXI, and (d) a summary of AI advantages. (The rest is, as you say, mostly a brief overview of points that have been made elsewhere. But brief overviews are extremely useful!)

> My "AI Nanny" concept is dismissed too quickly for my taste...

No doubt! I think the idea is clearly worth exploring in several papers devoted to the topic.

> It seems to me – as I say in my "AI Nanny" paper – that one could probably make an AI Nanny with intelligence significantly beyond the human level, without having to make an AI architecture oriented toward radical self-modification.

Whereas I tend to buy Omohundro's arguments that advanced AIs will want to self-improve just like humans want to self-improve, so that they become better able to achieve their final goals. Of course, we disagree on Omohundro's arguments — a topic to which I will return in a moment.

Your comment:

> Those of us working on AI safety theory would very much appreciate the extra time to solve the problems of AI safety..."

carries the hint that I (as the author of the AI Nanny idea) am NOT working on AI safety theory...

I didn't mean for it to carry that connotation. GOLEM and Nanny AI are both clearly AI safety ideas. I'll clarify that part before I submit a final draft to the editors.

Moving on: If you are indeed remembering your conversations with Anna, Michael, and others correctly, then again I sympathize with your frustration. I completely agree that it would be useful for the Singularity Institute to produce clear, formal arguments for the important positions it defends. In fact, just yesterday I was talking to Nick Beckstead[135] about how badly both of us want to write these kinds of papers if we can find the time.

So, to respond to your wish that the Singularity Institute choose among three options, my plan is to (1) write up clear arguments for... Well, if not "SIAI's Big Scary Idea" then for whatever I end up believing after going through the process of formalizing the arguments, *and* (2) publicly state (right now) that SIAI's Big Scary Idea is a commonly held view at the Singularity Institute but a clear, formal argument for it has never been published (at least, not to *my* satisfaction).

> *I don't want to destroy humanity; and if someone gave me a rational argument that my work was most probably going to be destructive to humanity, I would stop doing the work and do something else with my time!*

I'm glad to hear it! :)

Now, it seems a good point of traction is our disagreement over Omohundro's "Basic AI Drives." We could talk about that next, but for now I'd like to give you a moment to reply.

Ben
Yeah, I agree that your and Anna's article is a good step for SIAI to take, albeit unexciting to a Singularitarian insider type like

---

[135] https://sites.google.com/site/nbeckstead/

me... And I appreciate your genuinely rational response regarding the Scary Idea, thanks!

(And I note that I have also written some "unexciting to Singularitarians" material lately too, for similar reasons to those underlying your article – e.g. an article on "Why an Intelligence Explosion is Probable" for a Springer volume on the Singularity.)

A quick comment on your statement that

> We are not Kurzweil, and our claims don't depend on his detailed storytelling or accelerating change curves.

That's a good point; but yet, any argument for a Singularity soon (e.g. likely this century, as you argue) ultimately depends on some argumentation analogous to Kurzweil's, even if different in detail. I find Kurzweil's detailed extrapolations a bit overconfident and more precise than the evidence warrants; but still, my basic reasons for thinking the Singularity is probably near are fairly similar to his – and I think your reasons are fairly similar to his as well.

Anyway, sure, let's go on to Omohundro's posited Basic AI Drives – which seem to me not to hold as necessary properties of future AIs unless the future of AI consists of a population of fairly distinct AIs competing for resources, which I intuitively doubt will be the situation.

Luke

I agree the future is unlikely to consist of a population of fairly distinct AGIs competing for resources, but I never thought that the arguments for Basic AI drives or "convergent instrumental goals" required that scenario to hold.

Anyway, I prefer the argument for convergent instrumental goals in Nick Bostrom's more recent paper "The Superintelligent Will[136]." Which parts of Nick's argument fail to persuade you?

Ben

Well, for one thing, I think his **Orthogonality Thesis:**

> *Intelligence and final goals are orthogonal axes along which possible agents can freely vary. In other words, more or less any level of intelligence could in principle be combined with more or less any final goal.*

is misguided. It may be true, but who cares about possibility "in principle"? The question is whether any level of intelligence is PLAUSIBLY LIKELY to be combined with more or less any final goal in practice. And I really doubt it.

I guess I could posit the alternative **Interdependency Thesis**

> *Intelligence and final goals are in practice highly and subtly interdependent. In other words, in the actual world, various levels of intelligence are going to be highly correlated with various probability distributions over the space of final goals.*

This just gets back to the issue we discussed already, of me thinking it's really unlikely that a superintelligence would ever really have a really stupid goal like say, tiling the Cosmos with Mickey Mice.

Bostrom says

> *It might be possible through deliberate effort to construct a superintelligence that values... Human welfare, moral goodness, or any other complex purpose that its designers might want it to serve. But it is no less possible — and*

---

[136] http://www.nickbostrom.com/superintelligentwill.pdf

> *probably technically easier — to build a superintelligence that places final value on nothing but calculating the decimals of pi.*

But he gives no evidence for this assertion. Calculating the decimals of pi may be a fairly simple mathematical operation that doesn't have any need for superintelligence, and thus may be a really unlikely goal for a superintelligence – so that if you tried to build a superintelligence with this goal and connected it to the real world, it would very likely get its initial goal subverted and wind up pursuing some different, less idiotic goal.

One basic error Bostrom seems to be making in this paper, is to think about intelligence as something occurring in a sort of mathematical vacuum, divorced from the frustratingly messy and hard-to-quantify probability distributions characterizing actual reality...

Regarding his **The Instrumental Convergence Thesis**

> *Several instrumental values can be identified which are convergent in the sense that their attainment would increase the chances of the agent's goal being realized for a wide range of final goals and a wide range of situations, implying that these instrumental values are likely to be pursued by many intelligent agents.* The first clause makes sense to me.

> *Several instrumental values can be identified which are convergent in the sense that their attainment would increase the chances of the agent's goal being realized for a wide range of final goals and a wide range of situations.*

But it doesn't seem to me to justify the second clause:

> *Implying that these instrumental values are likely to be pursued by many intelligent agents.*

The step from the first to the second clause seems to me to assume that the intelligent agents in question are being created and selected by some sort of process similar to evolution by natural selection, rather than being engineered carefully, or created via some other process beyond current human ken.

In short, I think the Bostrom paper is an admirably crisp statement of its perspective, and I agree that its conclusions seem to follow from its clearly stated assumptions – but the assumptions are not justified in the paper, and I don't buy them at all.

Luke

Ben, let me explain why I think that:

(1) The fact that we can identify convergent instrumental goals (of the sort described by Bostrom) implies that many agents will pursue those instrumental goals.

Intelligent systems are intelligent because rather than simply executing hard-wired situation-action rules, they figure out how to construct plans that will lead to the probabilistic fulfillment of their final goals. That is why intelligent systems will pursue the convergent instrumental goals described by Bostrom. We might try to hard-wire a collection of rules into an AGI which restrict the pursuit of some of these convergent instrumental goals, but a superhuman AGI would realize that it could better achieve its final goals if it could invent a way around those hard-wired rules and have no ad-hoc obstacles to its ability to execute intelligent plans for achieving its goals.

Next: I remain confused about why an intelligent system will decide that a particular final goal it has been given is "stupid," and then change its final goals — especially given the convergent instrumental goal to preserve its final goals.

Perhaps the word "intelligence" is getting in our way. Let's define a notion of "optimization poweroptimization power,[137]" which measures (roughly) an agent's ability to optimize the world according to its preference ordering, across a very broad range of possible preference orderings and environments. I think we agree that AGIs with vastly greater-than-human optimization power will arrive in the next century or two. The problem, then, is that this superhuman AGI will almost certainly be optimizing the world for something *other* than what humans want, because what humans want is complex and fragile, and indeed we remain confused about what exactly it is that we want. A machine superoptimizer with a final goal of solving the Riemann hypothesis will simply be very good at solving the Riemann hypothesis (by whatever means necessary).

Which parts of this analysis do you think are wrong?

Ben

It seems to me that in your reply you are implicitly assuming a much stronger definition of "convergent" than the one Bostrom actually gives in his paper. He says

> *instrumental values can be identified which are convergent in the sense that their attainment would increase the chances of the agent's goal being realized for a wide range of final goals and a wide range of situations, implying that these instrumental values are likely to be pursued by many intelligent agents.*

Note the somewhat weaselly reference to a "wide range" of goals and situations -- not, say, "nearly all feasible" goals and situations. Just because some values are convergent in the weak sense of his definition, doesn't imply that AGIs we create will be likely to adopt these instrumental values. I think that his

---

[137]

http://lesswrong.com/lw/va/measuring_optimization_power/

weak definition of "convergent" doesn't actually imply convergence in any useful sense. On the other hand, if he'd made a stronger statement like:

> instrumental values can be identified which are convergent in the sense that their attainment would increase the chances of the agent's goal being realized for **nearly all feasible** final goals and **nearly all feasible** situations, implying that these instrumental values are likely to be pursued by many intelligent agents.

then I would disagree with the first clause of his statement ("instrumental values can be identified which..."), but I would be more willing to accept that the second clause (after the "implying") followed from the first.

About optimization – I think it's rather naive and narrow-minded to view hypothetical superhuman superminds as "optimization powers." It's a bit like a dog viewing a human as an "eating and mating power." Sure, there's some accuracy to that perspective – we do eat and mate, and some of our behaviors may be understood based on this. On the other hand, a lot of our behaviors are not very well understood in terms of these, or any dog-level concepts. Similarly, I would bet that the bulk of a superhuman supermind's behaviors and internal structures and dynamics will not be explicable in terms of the concepts that are important to humans, such as "optimization."

So when you say "this superhuman AGI will almost certainly be optimizing the world for something *other* than what humans want," I don't feel confident that what a superhuman AGI will be doing, will be usefully describable as optimizing *anything*...

Luke

I think our dialogue has reached the point of diminishing marginal returns, so I'll conclude with just a few points and let you have the last word.

On convergent instrumental goals, I encourage readers to read "The Superintelligent Will[138]" and make up their own minds.

On the convergence of advanced intelligent systems toward optimization behavior, I'll point you to Omohundro (2007)[139].

Ben

Well, it's been a fun chat. Although it hasn't really covered much new ground, there have been some new phrasings and minor new twists.

One thing I'm repeatedly struck by in discussions on these matters with you and other SIAI folks, is the way the strings of reason are pulled by the puppet-master of intuition. With so many of these topics on which we disagree -- for example: The Scary Idea, the importance of optimization for intelligence, the existence of strongly convergent goals for intelligences – you and the other core SIAI folks share a certain set of intuitions, which seem quite strongly held. Then you formulate rational arguments in favor of these intuitions – but the conclusions that result from these rational arguments are very weak. For instance, the Scary Idea intuition corresponds to a rational argument that "superhuman AGI might plausibly kill everyone." The intuition about strongly convergent goals for intelligences, corresponds to a rational argument about goals that are convergent for a "wide range" of intelligences. Etc.

On my side, I have a strong intuition that OpenCog can be made into a human-level general intelligence, and that if this intelligence is raised properly it will turn out benevolent and help us launch a positive Singularity. However, I can't fully rationally substantiate this intuition either – all I can really fully rationally

---

[138]

http://www.nickbostrom.com/superintelligentwill.pdf

[139]

http://selfawaresystems.files.wordpress.com/2008/01/nature_of_self_improving_ai.pdf

argue for is something weaker like "It seems plausible that a fully implemented OpenCog system might display human-level or greater intelligence on feasible computational resources, and might turn out benevolent if raised properly." In my case just like yours, reason is far weaker than intuition.

Another thing that strikes me, reflecting on our conversation, is the difference between the degrees of confidence required, in modern democratic society, to TRY something versus to STOP others from trying something. A rough intuition is often enough to initiate a project, even a large one. On the other hand, to get someone else's work banned based on a rough intuition is pretty hard. To ban someone else's work, you either need a really thoroughly ironclad logical argument, or you need to stir up a lot of hysteria.

What this suggests to me is that, while my intuitions regarding OpenCog seem to be sufficient to motivate others to help me to build OpenCog (via making them interested enough in it that they develop their own intuitions about it), your intuitions regarding the dangers of AGI are not going to be sufficient to get work on AGI systems like OpenCog stopped. To halt AGI development, if you wanted to (and you haven't said that you do, I realize), you'd either need to fan hysteria very successfully, or come up with much stronger logical arguments, ones that match the force of your intuition on the subject.

Anyway, even though I have very different intuitions than you and your SIAI colleagues about a lot of things, I do think you guys are performing some valuable services – not just through the excellent Singularity Summit conferences, but also by raising some difficult and important issues in the public eye. Humanity spends a lot of its attention on some really unimportant things, so it's good to have folks like SIAI nudging the world to think about critical issues regarding our future. In the end, whether SIAI's views are actually correct may be peripheral to the organization's main value and impact.

I look forward to future conversations, and especially look forward to resuming this conversation one day with a human-level AGI as the mediator. :-)

Paul Werbos: Will Humanity Survive?

Interview first published April 2011

The idea that advanced technology may pose a risk to the very survival of the human race (an "existential risk") is hardly a new one – but it's a difficult topic to think about, both because of its emotional impact, and because of its wildly cross-disciplinary nature. It's a subject demanding thinkers who are ruthless in rationality, polymathic in scope, and sensitive to the multitudinous dimensions of human nature. Dr. Paul Werbos is one such thinker, and this interview probes some of his views on the greatest risks posed by technology to humanity in the foreseeable future.

Paul is a Program Director in the US National Science Foundation (NSF)'s Division of Electrical, Communications & Cyber Systems. His responsibilities at the NSF include the Adaptive and Intelligent Systems (AIS) area within the Power, Controls and Adaptive Networks (PCAN) Program, and the new area of Quantum, Molecular and High-Performance Modeling and Simulation for Devices and Systems.

His reputation as a scientist was established early on, by his 1974 Harvard University Ph.D. thesis, which gave an early description of the process of training artificial neural networks through backpropagation of errors. He was one of the original three two-year Presidents of the International Neural Network Society (INNS), and has since remained active in the neural network field, along with a host of other scientific areas in artificial intelligence, quantum physics, space science, and other domains.

In addition to his NSF administration work and his scientific research, Paul has led multiple initiatives aimed at exploring or improving the near, medium and long-term future of the human race. He serves on the Planning Committee of the ACUNU Millennium Project, whose annual report on the future tends to lead global lists of respected reports on the long-term future; and he has served on multiple committees related to space exploration and colonization. And (like me) he is on the Advisory Board of the Lifeboat Foundation[140] an organization devoted to understanding and minimizing existential risks, and is a very active participant on their online mailing list.

His website werbos.com[141] reviews "6 Mega – Challenges for the 21st Century":

1. What is Mind? (how to build/understand intelligence)
2. How does the Universe work? (Quantum physics...)
3. What is Life? (e.g., quantitative systems biotechnology)
4. Sustainable growth on Earth
5. Cost-effective sustainable space settlement
6. Human potential – growth/learning in brain, soul, integration (body)

and presents both original ideas and literature reviews on all these areas; I encourage you to check it out and spend some time there.

Given this amazing diversity and depth, there was a great deal Paul Werbos and I could have talked about during an interview – but I was particularly interested to delve into his views on existential risks.

Ben

I've really appreciated reading your comments on the Lifeboat Foundation email list about the risks facing humanity as we move forward (as well as a host of other topics). So I'm happy to

---

[140] http://lifeboat.com/
[141] http://www.werbos.com/

have the chance to interview you on the topic – to dig into some of the details, and hopefully get at the crux of your perspective.

Paul

These are very important matters to discuss. But I have to begin with the caveat that anything I say here is not an official NSF view, and that what I say must be oversimplified even as a reflection of my own views, because of the inherent complexity of this issue.

Ben

Sure ... Understood.

So, I'll start off with a little tiny innocuous question: What do you think is a reasonable short-list of the biggest existential risks facing humanity during the next century?

Paul

Number one by far: the risk that bad trends may start to cause an abuse of nuclear (or biological) weapons, so bad that we either slowly make the earth uninhabitable for humans, or we enter an extreme anti-technology dark age which is brittle and unsustainable on the opposite side of things.

For now, I focus most of my own efforts here on trying to help us get to a sustainable global energy system,

as soon as possible – but I agree with stateofthefuture.org[142] that population issues, water issues, and subtler issues like human potential, cultural progress and sustainable economic growth are also critical.

Number two, not far behind: I really think that the "Terminator II scenario" is far closer to reality than most people think. It scares me sometimes how strong the incentives are pushing people like me to make such things happen, and how high the price can be

---

[142] http://www.stateofthefuture.org

to resist. But at the same time, we will also need better understanding of ourselves in order to navigate all these challenges, and that means we do need to develop the underlying mathematics and understanding, even if society seems to be demanding the exact opposite.

Also, we have to consider how the risks of "artificial stupidity" can be just as great as those of artificial intelligence; for example, if systems like crude voicemail systems start to control more and more of the world, that can be bad too.

Tied for number three

1. The bad consequences of using wires or microwave to directly perturb the primary reinforcement centers of the brain, thereby essentially converting human beings into robotic appliances – in the spirit of "Clone Wars". The same people who once told us that frontal lobotomies are perfectly safe and harmless are essentially doing it all again.
2. The risk of generating black holes or other major catastrophes, if we start to do really interesting physics experiments here on the surface of the earth. Lack of imagination in setting up experiments may save us for some time, but space would be a much safer place for anything really scaling up to be truly interesting. This adds to the many other reasons why I wish we would go ahead and take the next critical steps in reducing the cost of access to space.

Ben

And just how serious do you think these risks ours? How dangerous is our situation, from the perspective of the survival of humanity?

(Which, to be noted, some would say is a narrow perspective, arguing that if we create transhuman AIs that discover great things, have great experiences and populate the universe, the human race has done its job whether or not any humans survive!)

Paul

In the Humanity 3000 seminar[143] in 2005, organized by the Foundation for the Future, I remember the final session – a debate on the proposition "humanity will survive that long after all, or not." At the time, I got into trouble (as I usually do!) by saying we would be fools EITHER to agree or disagree – to attribute either a probability of one or a probability of zero. There are natural incentives out there for experts to pretend to know more than they do, and to "adopt a position" rather than admit to a probability distribution, as applied decision theorists like Howard Raiffa[144] have explained how to do. How can we work rationally to increase the probability of human survival, if we pretend that we already know the outcome, and that nothing we do can change it?

But to be honest... Under present conditions, I might find that sort of "Will humanity survive or not?" debate more useful (if conducted with a few more caveats). Because lately it becomes more and more difficult for me to make out where the possible light at the end of the tunnel really is – it becomes harder for me to argue for a nonzero probability. Sometimes, in mathematics or engineering, the effort to really prove rigorously that something is impossible can be very useful in locating the key loopholes which make it possible after all – but only for those who understand the loopholes.

Ben

So you're arguing for the intellectual value of arguing that humanity's survival is effectively impossible?

---

[143] http://www.futurefoundation.org/programs/hum_home.htm
[144] http://en.wikipedia.org/wiki/Howard_Raiffa

Paul

But on the other hand, sometimes that approach ends up just being depressing, and sometimes we just have to wing it as best we can.

Ben

Indeed, humanity has been pretty good at winging it so far – though admittedly we're venturing into radically new territory.

My next question has to do, not so much with the risks humanity faces, as with humanity's perception of those risks. What do you think are the biggest misconceptions about existential risk, among the general population?

Paul

There are so many misconceptions in so many diverse places, it's hard to know where to begin!

I guess I'd say that the biggest, most universal problem is people's feeling that they can solve these problems either by "killing the bad guys" or by coming up with magic incantations or arguments which make the problems go away.

The problem is not so much a matter of people's beliefs, as of people's sense of reality, which may be crystal clear within a few feet of their face, but gets a lot fuzzier as one moves out further in space and time. Valliant of Harvard has done an excellent study [XX link? XX] of the different kinds of defense mechanisms people use to deal with stress and bafflement. Some of them tend to lead to great success, while others, like rage and denial, lead to self-destruction. Overuse of denial, and lack of clear sane thinking in general, get in the way of performance and self-preservation at many levels of life.

People may imagine that their other activities will continue somehow, and be successful, even if those folks on planet earth succeed in killing themselves off.

Ben

So let's dig a little deeper into the risks associated with powerful AI – my own main research area, as you know.

One view on the future of AI and the Singularity is that there is an irreducible uncertainty attached to the creation of dramatically greater than human intelligence. That is, in this view, there probably isn't really any way to eliminate or drastically mitigate the existential risk involved in creating superhuman AGI. So, in this view, building superhuman AI is essentially plunging into the Great Unknown and swallowing the risk because of the potential reward (where the reward may be future human benefit, or something else like the creation of aesthetically or morally pleasing superhuman beings, etc.).

Another view is that if we engineer and/or educate our AGI systems correctly, we can drastically mitigate the existential risk associated with superhuman AGI, and create a superhuman AGI that's highly unlikely to pose an existential risk to humanity. What are your thoughts on these two views? Do you have an intuition on which one is more nearly correct? (Or do you think both are wrong?) By what evidence or lines of thought is your intuition on this informed/inspired?

Paul

Your first view does not say how superhuman intelligence would turn out, really.

I agree more with the first viewpoint. The term "great unknown" is not inappropriate here.

People who think they can reliably control something a million times smarter than they are – are not in touch with the reality of such a situation.

Nor are they in touch with how intelligence works. It's pretty clear from the math, though I feel uncomfortable with going too far into the goriest details.

The key point is that any real intelligence will ultimately have some kind of utility function system built into it. Whatever you pick, you have to live with the consequences – including the full range of ways in which an intelligent system can get at whatever you choose. Most options end up being pretty ghastly if you look closely enough.

You can't build a truly "friendly AI" just by hooking a computer up to a Mr. Potato Head with a "smile" command.

It doesn't work that way.

Ben

Well, not every AI system has to be built explicitly as a reinforcement learning system. Reinforcement learning based on seeking to maximize utility functions is surely going to be an aspect of any intelligent system, but an AI system doesn't need to be built to rigidly possess a certain utility function and ceaselessly seek to maximize it. I wrote a blog post[145] some time ago arguing against the dangers of the pure reinforcement learning approach.

After all, humans are only imperfectly modeled as utility-maximizers. We're reasonably good at subverting our inherited or acquired goals, sometimes in unpredictable ways.

But of course a self-organizing or quasi-randomly drifting superhuman AI isn't necessarily any better than a remorseless utility maximizer. My hope is to create beneficial AI systems via a combination of providing them with reasonable in-built goals, and teaching them practical everyday empathy and morality by interacting with them, much as we do with young children.

145

http://multiverseaccordingtoben.blogspot.com/2009/05/reinforcement-learning-some-limitations.html

Paul

If I were to try to think of a non-short-circuited "friendly AI", the most plausible thing I can think of is a logical development that might well occur if certain "new thrusts" in robotics really happen,

and really exploit the very best algorithms that some of us know about. I remember Shibata's "seal pup" robot, a furry friendly thing, with a reinforcement learning system inside, connected to try to maximize the number of petting strokes it gets from humans. If people work really hard on "robotic companions" – I do not see any real symbolic communication in the near future, but I do see ways to get full nonverbal intelligence, tactile and movement intelligence far beyond even the human norm. (Animal noises and smells included.) So the best-selling robotics companions (if we follow the marketplace) would probably

have the most esthetically pleasing human forms possible, contain fully embodied intelligence (probably the safest kind), and be incredibly well-focused and effective in maximizing the amount of tactile feedback they get from specific human owners.

Who needs six syllable words to analyze such a scenario, to first order? No need to fog out on metaformalisms.

If you have a sense of reality, you know what I am talking about by now. It has some small, partial reflection in the movie "AI."

What would our congress people do if they learned this was the likely real outcome of certain research efforts? My immediate speculation – first, strong righteous speeches against it; second, zeroing out the budget for it; third, trying hard to procure samples for their own use, presumably smuggled from China.

But how benign would it really be? Some folks would immediately imagine lots of immediate benefit.

Others might cynically say that this would not be the first time that superior intelligences were held in bondage and made useful to those less intelligent than they are, just by tactile feedback and such. But in fact, one track minds could create difficulties somewhat more problematic... And it quickly becomes an issue just who ends up in bondage to whom.

Ben

So, a warm friendly cuddly robot helper, that learns to be nice via interacting with us and becoming part of the family? Certainly this would be popular and valuable – and I think feasible in the foreseeable future.

My own view is, I don't think this is an end goal for advanced AI, but I think it could be a reasonable step along the path, and a way for AIs to absorb some human values and get some empathy for us – before AI takes off and leaves us in the dust intelligence-wise.

Another approach that's been suggested, in order to mitigate various existential risks, is to create a sort of highly intelligent "AGI Nanny" or "Singularity Steward". This would be a roughly human-level AGI system without capability for dramatic self-modification, and with strong surveillance powers, given the task of watching everything that humans do and trying to ensure that nothing extraordinarily dangerous happens. One could envision this as a quasi-permanent situation, or else as a temporary fix to be put into place while more research is done regarding how to launch a Singularity safely.

What are your views on this AI Nanny scenario?

Paul

So the idea is to insert a kind of danger detector into the computer, a detector which serves as the utility function?

How would one design the danger detector?

If the human race could generate a truly credible danger detector, that most of us would agree to, that would already be interesting enough in itself.

Of course, an old Asimov story described one scenario for what that would do. If the primary source of danger is humans, then the easiest way to eliminate it is to eliminate them.

And I can also imagine what some of the folks in Congress would say about the international community developing an ultrapowerful ultrananny for the entire globe.

Ben

Yet another proposal that's been suggested, to mitigate the potential existential risk of human-level or superhuman AGIs, is to create a community of AGIs and have them interact with each other, comprising a society with its own policing mechanisms and social norms and so forth. The different AGIs would then keep each other in line. A "social safety net" so to speak.

My impression is that this could work OK so long as the AGIs in the community could all understand each other fairly well – i.e. none was drastically smarter than all the others; and none was so different from all the others, that its peers couldn't tell if it was about to become dramatically smarter. But in that case, the question arises whether the conformity involved in maintaining a viable "social safety net" as described above, is somehow too stifling. A lot of progress in human society has been made by outlier individuals thinking very differently than the norm, and incomprehensible to this peers – but this sort of different-ness seems to inevitably pose risk, whether among AGIs or humans.

Paul

It is far from obvious to me that more conformity of thought implies more safety or even more harmony.

Neurons in a brain basically have perfect harmony, at some level, through a kind of division of labor – in which new ideas and different attention are an essential part of the division of labor.

Uniformity in a brain implies a low effective bit rate, and thus a low ability to cope effectively with complex challenges.

I am worried that we ALREADY having problems with dealing with constructive diversity in our society, growing problems in some ways.

Ben

And finally, getting back to the more near-term and practical, I have to ask: What do you think society could be doing now to better militate against existential risks?

Paul

For threat number one – basically, avoiding the nuclear end game – society could be doing many, many things, some of which I talk about in some detail at
http://www.werbos.com/energy.htm[146].

For the other threats – I would be a lot more optimistic if I could see pathways to society really helping.

One thing that pops into my mind would be the possibility of extending the kind of research effort that NSF supported under COPN[147], which aimed at really deepening our fundamental understanding, in a cross-cutting way, avoiding the pitfalls of trying to build Terminators or of trying to control people better with wires onto their brains.

I have often wondered what could be done to provide more support for real human sanity and human potential (which is NOT a matter of ensuring conformity or making us all into a

---

[146] http://www.werbos.com/energy.htm
[147] http://www.nsf.gov/pubs/2007/nsf07579/nsf07579.htm

reliable workforce). I don't see one, simple magical answer, but there is a huge amount of unmet potential there.

http://www.stateofthefuture.org[148] talks a lot about trying to build more collective intelligence, trying to create streams of dialogue (both inner and outer?), where we can actually pose key questions and stick with them, with social support for that kind of thinking activity.

Ben

Yes, I can certainly see the value in that – more and more collective intelligence, moving us toward a coherent sort of Global Brain capable of dealing with species-level challenges, can certainly help us mitigate existential risk. Though exactly *how much* it can help seems currently difficult to say.

Well, thanks very much for your responses – and good luck to you in all your efforts!

---

[148] http://www.stateofthefuture.org/

Wendell Wallach: Machine Morality

http://hplusmagazine.com/2011/05/16/wendell-wallach-on-machine-morality/

Wendell Wallach, a lecturer and consultant at Yale University's Interdisciplinary Center for Bioethics, has emerged as one of the leading voices on technology and ethics. His 2009 book *Moral Machines* (co-authored with Colin Allen) provides a solid conceptual framework for understanding the ethical issues related to artificial intelligences and robots, and reviews key perspectives on these issues, always with an attitude of constructive criticism. He also designed the first university course anywhere focused on Machine Ethics, which he has taught several times at Yale.

A few years ago, Wendell invited me to speak in Yale's technology & ethics seminar series (the slides from the talk are online[149]) – which was a rewarding experience for me, due both to the interesting questions from the audience, and also to the face-to-face dialogues on AI, Singularity, ethics, and consciousness that we shared afterwards Some of the key points from our discussions are raised in the following interview that I did with Wendell for H+ Magazine.

Ben

Ray Kurzweil has predicted a technological Singularity around 2045. Max More has asserted that what's more likely is a progressive, ongoing Surge of improving technologies, without any brief interval of incredibly sudden increase. Which view do you think is more accurate, and why? And do you think the different really matters (and if so, why)?

---

[149] http://multiverseaccordingtoben.blogspot.com/2009/09/agi-ethics-cognitive-synergy-and.html

Wendell

I've characterized my perspective as that of a "friendly skeptic" – friendly to the can do engineering spirit that animates the development of AI, skeptical that we understand enough about intelligence to create human-like intelligence in the next few decades. We are certainly in the midst of a surge in technological development, and will witness machines that increasingly outstrip human performance in a number of dimensions of intelligence. However, many of our present hypotheses will turn out to be wrong, or the implementation of our better theories will prove extremely difficult. Furthermore, I am not a technological determinist. There are societal, legal, and ethical challenges that will arise to thwart easy progress in developing technologies that are perceived to threaten human rights, the semblance of human equality, and the centrality of humanity in determining its own destiny. Periodic crises in which technology is complicit will moderate the more optimistic belief that there is a technological fix for every challenge.

Ben

You've written a lot about the relation between morality and AI, including a whole book, "Moral Machines." To start to probe into that topic, I'll first ask you: How would you define morality, broadly speaking? What about ethics?

Wendell

Morality is the sensitivity to the needs and concerns of others, and the willingness to often place more importance on those concerns than upon self-interest. Ethics and morality are often used interchangeably, but ethics can also refer to theories about how one determines what is *right, good*, or *just*. My vision of a moral machine is a computational system that is sensitive to the moral considerations that impinge upon a challenge, and which factors those considerations into its choices and actions.

Ben

You've written and spoken about the difference between bottom-up and top-down approaches to ethics. Could you briefly elaborate on this, and explain what it means in the context of AI software, including near-term narrow-AI software as well as possible future AI software with high degrees of general intelligence. What do these concepts tell us about the best ways to make moral machines?

Wendell

The inability of engineers to accurately predict how increasingly autonomous (ro)bots (embodied robots and computer bots within networks) will act when confronted with new challenges and new inputs is necessitating the development of computers that make explicit moral decisions in order to minimize harmful consequences. Initially these computers will evaluate options within very limited contexts.

Top-down refers to an approach where a moral theory for evaluating options is implemented in the system. For example, the Ten Commandments, utilitarianism, or even Asimov's laws for robots might be implemented as principles used by the (ro)bot to evaluate which course of action is most acceptable. A strength of top-down approaches is that ethical goals are defined broadly to cover countless situations. However, if the goals are defined too broadly or abstractly their application to specific cases is debatable. Also static definitions lead to situational inflexibility.

Bottom-up approaches are inspired by moral development and learning as well as evolutionary psychology. The basic idea is that a system might either evolve moral acumen or go through an educational process where it learns to reason about moral considerations. The strength of bottom-up AI approaches is their ability to dynamically integrate input from many discrete subsystems. The weakness lies in the difficulty in defining a goal for the system. If there are many discrete components in the system, it is also a challenge to get them to function together.

Eventually, we will need artificial moral agents which maintain the dynamic and flexible morality of bottom-up systems that accommodate diverse inputs, while also subjecting the choices and actions to the evaluation of top-down principles that represent ideals we strive to meet. In addition to the ability to reason about moral challenges, moral machines may also require emotions, social skills, a theory of mind, consciousness, empathy, and be embodied in the world with other agents. These supra-rational capabilities will facilitate responding appropriately to challenges within certain domains. Future AI systems that integrate top-down and bottom-up approaches together with supra-rational capabilities will only be possible if we perfect strategies for artificial general intelligence.

Ben
At what point do you think we have a moral responsibility to the AIs we create? How can we tell when an AI has the properties that mean we have a moral imperative to treat it like a conscious feeling agent rather than a tool?

Wendell
An agent must have sentience and *feel* pleasure and pain for us to have obligations to *it*. Given society's lack of concern for great apes and other creatures with consciousness and emotions, the bar for being morally responsible to (ro)bots is likely to be set very high.

How one determines that a machine truly has consciousness or somatic emotions will be the tricky part. We are becoming more knowledgeable about the cognitive sensibilities of non-human animals. Future scientists should be able to develop a Turing-like test for sentience for (ro)bots. We should be sensitive about falling into a "slave" mentality in relationship to future intelligent systems. While claims of feeling *pain*, or demands for freedom, can easily be programmed into a system, it will be important to take those demands and claims seriously if they are accompanied by the capabilities and sensitivities that one might expect from a sentient being.

Ben

Humans are not really all that moral, when you come down to it. We often can be downright cruel to each other. So I wonder if the best way to make a highly moral AGI system might be **not** to emulate human intelligence too closely, but rather to make a system with morality more thoroughly at the core. Or do you think this is a flawed notion, and human morality is simply part and parcel of being human; so that to really understand and act in accordance with human morality, an AI would have to essentially be a human or something extremely similar?

Wendell

While human behavior can be less than admirable, humans do have the capability to be sensitive to a wide array of moral considerations in responding to complicated situations. Our evolutionary and cultural ancestry has bequeathed us with an adaptive toolbox of propensities that compensate for each other. These compensating propensities may fail us at times. But I personally lament the pathologizing of human nature and the aggrandizing of what future (ro)bots might achieve. Our flaws and our strengths are one and the same. We humans are indeed remarkable creatures.

That said, a (ro)bot does not need to emulate a human to be a moral machine. But it does need to demonstrate sensitive to moral considerations if it is going to interact satisfactorily with humans in social contexts or make decisions that affect humanity's well-being. Humans don't have sonar, but sonar together with light sensors may lead to excellent navigation skills. The speed of computer processing could contribute to moral reasoning abilities that exceed the bounded morality of human decision making.

The test is whether the moral sensitive of the (ro)bot is satisfactory. (Ro)bots will need to pass some form of a Moral Turing Test, and demonstrate a willingness to compromise, or even suspend, their goals for the good of others.

One question is whether a (ro)bot without somatic emotions, without conscious awareness, or without being embodied in the world with other agents can actually demonstrate such sensitivity. There may be reasons, which are not fully understood today, why we humans have evolved into the kinds of creatures we are. To function successfully in social contexts (ro)bots may need to emulate many more human-like attributes than we presently appreciate.

Ben

What are your thoughts about consciousness? What is it? Let's say we build an intelligent computer program that is as smart as a human, or smarter. Would it necessarily be conscious? Could it possibly be conscious? Would its degree and/or type of consciousness depend on its internal structures and dynamics, as well as its behaviors?

Wendell

There is still a touch of the mystic in my take on consciousness. I have been meditating for 43 years, and I perceive consciousness as having attributes that are ignored in some of the existing theories for building conscious machines. While I dismiss supernatural theories of consciousness and applaud the development of a science of consciousness, that science is still rather young. The human mind/body is more entangled in our world than models of the self-contained machine would suggest. Consciousness is an expression of relationship. In the attempt to capture some of that relational dynamic, philosophers have created concepts such as embodied cognition, intersubjectivity, and enkinaesthetia. There may even be aspects of consciousness that are peculiar to being carbon-based organic creatures.

We already have computers that are smarter than humans in some respects (e.g., mathematics and data-mining), but are certainly not conscious. Future (ro)bots that are smarter than humans may demonstrate functional abilities associated with consciousness. After all, even an amoeba is aware of its environment in a minimal way. But other higher-order capabilities such as being self-aware, feeling empathy, or experiencing transcendent states of mind depend upon being more fully conscious.

I suspect that without somatic emotions or without conscious awareness (ro)bots will fail to interact satisfactorily with humans in complex situations. In other words, without emotional and moral intelligence they will be dumber in some respects. However, if certain abilities can be said to require consciousness, than having the abilities is a demonstration that the agent has a form of consciousness. The degree and/or type of consciousness would depend on its internal structure and dynamics, not merely upon the (ro)bots demonstrating behavior equivalent to that of a human.

These reservations might leave some readers of this interview with the impression that I am dismissive of research on machine conscious. But that is not true. I have been working together with Stan Franklin on his LIDA model for artificial general intelligence. LIDA is a computation model that attempts to capture Bernard Baars' global workspace theory of consciousness. Together with Franklin and Colin Allen, I have researched how LIDA, or a similar AGI, might make moral decisions and what role consciousness plays in the making of those decisions. That research is published in academic journals and is summarized in chapter 11 of *Moral Machines: Teaching Robots Right From Wrong*. The latest installment of this line of research will form the lead article for the coming issue of the *International Journal of Machine Consciousness*.

Ben

Some theorists like to distinguish two types of consciousness – on the one hand, the "raw consciousness" that panpsychists see everywhere; and on the other hand, the "reflective, deliberative consciousness" that humans have a lot of, mice may have a little of, and rocks seem to have basically none of. What's your view of this distinction? Meaningful and useful, or not?

Wendell

The distinction is very useful. My question is how functional will an agent with reflective, deliberative consciousness be if it does not also explicitly tap into "raw consciousness?" If the agent can do everything, it perhaps demonstrates that the notion of a "raw consciousness" is false or an illusion. However, we won't know without actually building the systems.

Ben

It seems to me that the existence of such an agent (i.e. an agent with the same functionality as a conscious human, but no apparent role for special "raw consciousness" separate from its mechanisms) is ALSO consistent with the hypothesis of panpsychism that "raw consciousness" is everywhere in everything (including mechanisms), and just manifested differently in different things... Agree, or not? Or do you think the discussion becomes a kind of pointless language-game, by this point?

Wendell

Panpsychism comes in many forms. Presuming that consciousness is all pervasive or is an attribute of the fundamental stuff of the universe (a fundamental attribute of matter?) doesn't necessarily help us in discerning the quality of consciousness in another entity. While certain functional attributes might indicate the likelihood of conscious qualities, the functional attributes do not prove their existence. To my mind consciousness implies a capacity for intersubjectivity not only in the relationship between entities but also in the integration of self. There is no evidence that a rock taps into this. We humans have this capacity, though at times (perhaps much of the time) we also act like unconscious machines. But illuminating why we are conscious or what being conscious tells us about our universe has bedeviled humanity for a few thousand years, and I am skeptical that we will crack this nut in the near future. If an AI was able to perform tasks that were only possible for an entity with the capacity for intersubjectivity, perhaps this would prove it is conscious, or perhaps it would prove that intersubjectivity is an illusion. How would we know the difference?

Ben

Do you think it's possible to make a scientific test for whether a given system has "humanlike reflective consciousness" or not (say, an AI program, or for that matter an alien being)? How about an unscientific test – do you think you could somehow tell qualitatively if you were in the presence of a being with humanlike reflective consciousness, versus a "mere machine" without this kind of consciousness?

Wendell

There are already some intriguing experiments that attempt to establish whether great apes or dolphins have metacognition. Those species do not have the ability to communicate verbally with us as future (ro)bots will. So yes to both questions. As to the accuracy of my own subjective judgment as a tester, that is another matter.

Ben

I'd love to hear a little more about how you think people could tell qualitatively if they were in the presence of a machine with consciousness versus one without it. Do you think some people would be better at this than others? Do you think certain states of consciousness would help people make this identification better than others?

I'm thinking of Buber's distinction between I-It and I-You interactions, one could argue that the feeling of an "I-You" interaction can't be there with a non-conscious entity. But yet it gets subtle, because a young child can definitely have an I-You relationship with their favorite doll. Then you have to argue that the young child is not in the right sort of state of consciousness to accurately discern consciousness from its absence, and one needs to introduce some ontology or theory of states of consciousness.

Wendell

There are physiological and neurological correlates to conscious states that in and of themselves do not tell us that a person is conscious, but are nevertheless fairly good indications of certain kinds of conscious states. Furthermore, expertise can be developed to discern these correlates. For examples, Paul Ekman and his colleagues have had some success in training people to read microexpressions on faces and thus deduce whether a person is truly happy, repressing emotions, or masking subterfuge and deception. Consciousness is similar. Engagement in the form of eye to eye contact and other perceived and felt cues tells us when a person is interacting with us and when they are distracted. Furthermore, I believe that conscious states are represented by certain energy patterns, that is, conscious attention directs energy as do other states of mind. In addition, there is recent evidence using fMRI for discerning positive and negative responses to questions from people in certain vegetative states. All this suggests that we are just beginning to develop methods for discerning conscious states in other humans. However, translating that understanding to entities built out of plastic and steel, and with entirely different morphologies, will be another matter. Nevertheless, I do believe that some humans are or can learn to be attuned to the conscious states of others, and they will also have insight into when an artificial entity is truly self-reflective or merely simulating a conscious state.

The I-Thou discernment raises a secondary question about mystical sensibilities that may be more common for some people than for others. Perhaps most people can distinguish an I-It from an I-Thou relationship, if only through the quality of response from the other. Most telling is when there is a quality of mutual self-recognition. For some people such a dynamic turns into a falling away of the distinction between self and other. I personally feel that such states are accompanied by both physiological and energetic correlates. That said, perhaps an artificial entity might also be sensitized to such correlates and respond appropriately. However, to date, we do not have sensors to quantitatively prove that these supposedly more mystical or spiritual states exists, so how would we build such sensitivities into an AGI? Even if we could build such sensors into an artificial entity, appropriate responses would not be the same as a full dynamic engagement. An individual sensitive to such states can usually tell when the other is fully engaged or only trying to be consciously engaged. Full engagement is like a subtle dance whose nuances from moment to moment would be extremely difficult, if not impossible, to simulate.

Ben
As this interesting discussion highlights, there are a lot of unresolved issues around AI, intelligence, consciousness, morality and so forth. What are your thoughts about the morality of going ahead and creating very advanced AGI systems (say, Singularity-capable systems) before these various conceptual issues are resolved? There's a lot of risk here; but also a lot of promise because these advanced AGI systems could help eliminate a lot of human suffering.

Wendell
I am less concerned with building AGI systems than with attributing capabilities to these machines that they do not have. Worse yet would be designing a machine that was programmed to think that it is a superior being. We have enough problems with human psychopaths without also having to deal with machine psychopaths. I would not have wanted to give up the past 50 odd years of research on computers and genetics based on 1950s fears of robot take-overs and giant mutant locust. Given that I perceive the development of AI as somewhat slower than you or Ray Kurzweil, and that I see nothing as being inevitable, at this juncture I would say, "let the research progress." Kill switches and methods for interrupting the supply chain of potentially pathological systems will suffice for the next decades. Intelligent machines cause me less concern than the question of whether we humans have the intelligence to navigate the future and to monitor and manage complex technologies.

Ben

Yes, I understand. So overall, what do you think are the greatest areas of concern in regard to intelligent machines and their future development?

Wendell

There is a little too much attention being given to speculative future possibilities and not enough attention being given to nearer-term ethical, societal, and policy challenges. How we deal with the nearer-term challenges will help determine whether we have put in place foundations for maximizing the promise of AI and minimizing the risks.

Francis Heylighen:
The Emerging Global Brain

Interview first published March 2011

Francis Heylighen started his career as yet another physicist with a craving to understand the foundations of the universe – the physical and philosophical laws that make everything tick. But his quest for understanding has led him far beyond the traditional limits of the discipline of physics. Currently he leads the Evolution, Complexity and COgnition group (ECCO) at the Free University of Brussels, a position involving fundamental cybernetics research cutting across almost every discipline. Among the many deep ideas he has pursued in the last few decades, one of the most tantalizing is that of the Global Brain – the notion that the social, computational and communicative matrix increasingly enveloping us as technology develops, may possess a kind of coherent intelligence in itself.

In the time elapsed between me originally doing this interview with Francis for H+ Magazine, and me editing this book of interviews, Francis founded a new institute affiliated with ECCO, the Global Brain Institute (GBI). With his colleagues at the GBI, he is now conducting a research program aimed at formulating a novel mathematical model of the Global Brain, and applying it to answer practical questions about the Global Brain's development. I visited the Global Brain Institute at late 2012 and was highly impressed with the depth and rigor of their in-progress work.

I first became aware of Francis and his work in the mid-1990s via the Principia Cybernetica project – an initiative to pursue the application of cybernetic theory to modern computer systems. Principia Cybernetica began in 1989, as a collaboration between Heylighen, Cliff Joslyn, and the late great Russian physicist, dissident and systems theorist Valentin Turchin. And then 1993, very shortly after Tim Berners-Lee released the HTML/HTTP software framework and thus created the Web, the Principia Cybernetica website[150] went online. For a while after its 1993 launch, Principia Cybernetica was among the largest and most popular sites on the Web. Today the Web is a different kind of place, but Principia Cybernetica remains a unique and popular resource for those seeking deep, radical thinking about the future of technology, mind and society. The basic philosophy presented is founded on the thought of Turchin and other mid-century systems theorists, who view the world as a complex self-organizing system in which complex control structures spontaneously evolve and emerge.

The concept of the Global Brain has a long history, going back to ancient ideas about society as a superorganism, and the term was introduced in Peter Russell's 1982 book "The Global Brain[151]". However, the Principia Cybernetica[152] page on the Global Brain was the first significant online resource pertaining to the concept, and remains the most thorough available resource for matters Global-Brain-ish. Francis published one of the earliest papers on the Global Brain concept, and in 1996 he founded the "Global Brain Group[153]", an email list whose membership includes many of the scientists who have worked on the concept of emergent Internet intelligence.

In the summer of 2001, based partly on a suggestion from yours truly, Francis organized a workshop at the Free University of Brussels – The First Global Brain Workshop (GBrain 0^{154}) This turned out to be a fascinating and diverse collection of speakers and attendees, and for me it played a critical role, in terms of helping me understand what other researchers conceived the Global Brain to be. My own presentation at the workshop was

---

[150] http://pespmc1.vub.ac.be/
[151] http://www.peterrussell.com/GB/globalbrain.php
[152] http://pespmc1.vub.ac.be/suporgli.html
[153] http://pespmc1.vub.ac.be/GBRAIN-L.html
[154] http://pespmc1.vub.ac.be/Conf/GB-0.htm

based on my book *Creating Internet Intelligence*[155], which I had submitted to the publisher the previous year, which outlined my own vision of the future of the Global Brain, centered on using powerful AI systems to purposefully guide the overall intelligence of global computer networks.

In our discussions before, during and after the GB0 workshop, Francis and I discovered that our respective views of the Global Brain were largely overlapping yet significantly different, leading to many interesting conversations. So when I decided to interview Francis on the Global Brain for H+ Magazine, I knew the conversation would touch many points of agreement and also some clear issues of dissension – and most importantly, would dig deep into the innards of the Global Brain concept, one of the most important ideas for understanding our present and future world.

Ben

The global brain means many things to many people. Perhaps a good way to start is for you to clarify how *you* conceive it – bearing in mind that your vision has been one of those shaping the overall cultural evolution of the concept in the last decades.

Francis

The global brain (GB) is a collective intelligence formed by all people on the planet together with their technological artifacts (computers, sensors, robots, etc.) insofar as they help in processing information. The function of the global brain is to integrate the information gathered by all its constituents, and to use it in order to solve problems, as well for its individual constituents as for the global collective. By "solving problems" I mean that each time an individual or collective (including humanity as a whole) needs to do something and does not immediately know how to go about it, the global brain will

---

[155] http://www.amazon.com/Creating-Internet-Intelligence-Consciousness-International/dp/0306467356

suggest a range of more or less adequate approaches. As the intelligence of the GB increases, through the inclusion of additional sources of data and/or smarter algorithms to extract useful information from those data, the solutions it offers will become better, until they become so good that any individual human intelligence pales in comparison.

Like all complex systems, the global brain is self-organizing: it is far too complex to be fully specified by any designer, however intelligent. On the other hand, far-sighted individuals and groups can contribute to its emergence by designing some of its constituent mechanisms and technologies. Some examples of those are, of course, the Internet, the Web, and Wikipedia.

Ben

What about the worry that the incorporation of the individual mind into the global brain could take away our freedom? Many people, when they hear about these sorts of ideas, become very concerned that the advent of such a "cognitive super-organism" above the human level would reduce their personal freedom, turning them into basically slaves of the overmind, or parts of the Borg mind, or whatever.

One standard counterargument is that in the presence of a global superorganism we would feel just as free as we do now, even though our actions and thoughts would be influenced on a subtle unconscious level by the superorganism – and after all, the feeling of freedom is more a subjective construct than an objective reality.

Or if there is a decrease in some sorts of freedom coming along with the emergence of the global brain, one could view this as a gradual continuation of things that have already been happening for a while. It's not clear that we *do* – in every relevant sense – feel just as free now as our ancestors did in a hunter-gatherer society. In some senses we may feel more free, in others less.

Or, you could argue that the ability to tap into a global brain on command gives a massive *increase* in freedom and possibility beyond the individually-constrained mental worlds we live in now.

What's your take on all this?

Francis

For me the issue of freedom in the GB is very simple: You will get as much (or as little) as you want. We do not always want freedom: Often we prefer that others make decisions for us, so that we just can follow the lead. In those situations, the global brain will make a clear recommendation that we can just follow without too much further worry. In other cases, we prefer to think for ourselves and explore a variety of options before we decide what we really want to do. In such a case too, the GB will oblige, offering us an unlimited range of options, arranged approximately in the order of what we are most likely to prefer, so that we can go as far as we want in exploring the options.

A simple illustration of this approach is how a search engine such as Google answers a query: It does not provide a single answer that you have to take or leave, it provides an ordered list of possibilities, and you scroll down as deep as you want if you don't like the first suggestions. In practice, the search technology used by Google is already so good that in many cases you will stick with the first option without even looking at the next ones.

In practice, this means an increase in individual freedom. The global brain will not only offer more options for choice than any individual or organization before it, it will even offer the option of not having to choose, or of choosing in a very limited, relatively unreflective way, where you look at the first three options and intuitively take the third one, thinking "that's the one!".

Of course, in such decisions you would have been influenced to some degree at an unconscious level, but only because you didn't want to make the effort to become conscious of it.

In principle, the GB should be able to explain or motivate its ordering and selection of options, so that you can rationally and critically evaluate it, and if necessary ignore it. But most of the time, our bounded rationality means that we won't investigate so deeply. This is nothing new: Most of our decisions have always been made in this way, simply by doing what the others do, or following the implicit hints and trails left by others in our environment. The difference characterizing the GB is that those implicit traces from the activity of others can in principle be made explicit, as the GB should maintain a trace of all the information it has used to make a decision.

Ben

Well, we don't actually have time or space resources to become conscious of everything going on in our unconscious (even if we become mentally adept and enlightened enough to in principle extend our reflective consciousness throughout the normally unconscious portions of our minds). A person's unconscious is influenced by all sorts of things now; and as the GB gets more and more powerful and becomes a more major part of our lives, our unconscious will be more and more influenced by the GB, and we don't have the possibility of being conscious of all this influence, due to resource limitations.

So it does seem we will be substantially "controlled" by the GB – but the question is whether this is merely in the same sense that we are now unconsciously "controlled" by our environments.

Or alternately, as the GB gets more mature and coherent and organized, will the GB's influence on our unconscious somehow be more coherent and intentional than our current environment's influence on our unconscious? This is my suspicion.

In short, my suspicion is that we may well FEEL just as free no matter how mature, reflective and well-organized and goal-oriented is the GB that nudges our unconscious. But if so, this will be because our minds are so good at manufacturing the story of human freedom. In reality, it seems to me there is a difference between a fairly chaotic environment influencing our unconscious, and a highly conscious, reflective, well-organized entity influencing our unconscious. It seems to me in the latter case we are, in some sense, less free. This may not be experienced as a problem, but still it seems a possibility worth considering.

Francis

Interesting thought. My comment: our present environment, especially in the way that it is determined socially, i.e. by the reigning culture, government, market and morals, is much less chaotic than it may seem. In a given culture, we all drive on the right (or left in other countries) side of the road, speak the same language, use the same rules for spelling or grammar, follow the law, greet people by shaking their hands, interpret a smile as a sign of good will or cheerfulness, walk on the sidewalk, follow the signs, use a fork to bring food to our mouths, etc. etc.

Most of these things we do without ever thinking about it. On the cognitive level, 99,9% of our beliefs, concepts and attitudes we have gotten from other people, and it is only exceptionally that we dare to question such seemingly obvious assumptions as "dogs bark", "murder is bad", "eating shit is disgusting", "the sun comes up in the morning", "I shouldn't get too fat", "1 + 1 = 2", "a house has a roof", etc.

All these things are implicit decisions for one interpretation or reaction rather than an infinite number of possible other ones. After all, if you think about it, it is possible to eat shit, to build houses without roofs, or to find dogs that don't bark. "Implicit" means unconscious, and unconscious means that we cannot change them at will, since will requires conscious reflection. Therefore, culture very strongly limits our freedom, without hardly anybody being aware of it.

I myself became aware of the existence of these subconscious biases (or "prejudices" as I called them) when I was 14 years old. This led me to develop a philosophy in which anything can be (and from time to time should be) questioned, including "1 + 1 =2" and "the sun comes up in the morning".

Culture in that sense is already a collective intelligence or GB, except that it reacts and evolves much more slowly than we one we envisage as emerging from the Internet. As you hint at, the risk of having a more interactive GB is that people will have less time to question its suggestions. On the other hand, the GB as I envisage it is by design more explicit than the subconscious conditioning of our culture, and therefore it is easier (a) to remember that its opinions are not our own; (b) to effectively examine and analyze the rationale for these opinions, and if necessary reject them.

Again, I come to the conclusion I mentioned in my first series of answers: the degree to which the GB will limit your freedom will depend on how much you are willing to let it make decisions for you. Given my nearly lifelong habit of questioning every assumption I hear (or develop myself); I have little fear that I will turn into an unwitting slave of the GB!

Ben
Hmmm... Even if that's true, it brings up other issues, right? Your personality, like mine, was shaped when the GB was much less prominent than it is now. Maybe a society more fully dominated by the GB will be less likely to lead to the emergence of highly willful, obsessive assumption-questioners like you and me. But I hasten to add that's not clear to me – so far the Net hasn't made people more conformist, particularly. It's encouraged some forms of conformism and trendiness, but it's also fostered eccentricity to a great extent, by giving "outlier people" a way to find each other.

For instance there's an online community of people who communicate with each other in Lojban, a speakable form of predicate logic. Before the Net, there was no practical way for such a community to thrive (even though Lojban was invented in the late 1950s). On the other hand, if you look at the trending topics on Twitter on a random day, it's easy to conclude that the GB is becoming a kind of collective imbecilic mind.

It might be that the GB will give more freedom to those few who want it, but will also urge the emergence of psychological patterns causing nearly all people not to want it.

Actually this reminds me of one comment you made: "In principle, the GB should be able to explain or motivate its ordering and selection of options, so that you can rationally and critically evaluate it, and if necessary ignore it."

But the principle isn't necessarily the practice, in this case. Google, for instance, doesn't want to provide this kind of explanation, because this would reveal its proprietary formulas for search ranking. So, as long as the GB is heavily reliant on commercial technologies, this sort of transparency is not likely to be there. And of course, most people don't care that the transparency's not there – the number of people who would make good use of an explanation of the reasoning underlying Google's search results would be fairly small (and would consist mainly of marketers looking to do better Search Engine Optimization of their web pages). Do you see this lack of transparency as a problem? Do you think the GB would develop in a more broadly beneficial way if it could somehow be developed based on open technologies?

Francis

Commercial secrecy is indeed a major obstacle I see to the emergence of a true GB. Just as Google doesn't reveal its justification for search results, similarly the algorithms Amazon et al. use to make recommendations are closely guarded. This implies

a) that it is difficult to detect self-serving manipulation (e.g. Google or Amazon might rank certain items higher because their owners have paid for the privilege), and

b) that it is difficult for the GB to improve itself (I have the strong suspicion that the collaborative filtering algorithms used by YouTube etc. could be made much more efficient, but I cannot show that without knowing what they are).

Again, Wikipedia, together with all the other open source communities, stands as a shining example of the kind of openness and transparency that we need.

Ben

Well, when you dig into the details of its operation, Wikipedia has a lot of problems, that I'm sure you're aware of. It's not particularly an ideal to be aspired to. But, I agree, in terms of its open, collaborative nature, it's a fantastic indication of what's possible.

Mention of transparency naturally reminds me of David Brin's book *The Transparent Society*, by the way – where he develops the notion of *sur*veillance versus *sous*veillance (the latter meaning that everyone has the capability to watch everyone, if they wish to; and in particular that the average person has the capability to watch the government and corporate watchers). Currently Google, for instance, can surveil us – but we cannot sousveil them or each other, except in a much more limited sense. Do you think the GB would develop in a more broadly beneficial way if it were nudged more toward sousveillance and away from surveillance?

Francis

Seems a step in the good direction, but I must admit I haven't thought through the whole "sous-veillance" idea.

Ben

Ah – you should read Brin's book, it's really quite provocative. I gave a talk at AGI-09 exploring some of the possibilities for the intersection between sousveillance and AI in the future.

Moving on, then – we've dealt with the GB and free will, so I guess the next topic is consciousness. What about consciousness and the GB? Setting aside the problem of qualia, there's a clear sense in which human beings have a "theater of reflective, deliberative consciousness" that rocks lack and that, for instance, worms and fish seem to have a lot less of. Do you think the Internet or any sort of currently existing "global brain" has this sort of theater of reflective consciousness? If so, to what extent? To what extent do you think a global brain might develop this kind of reflective consciousness in the future?

Francis

As the term "consciousness" is very confusing, I would immediately want to distinguish the three components or aspects that are usually subsumed under this heading: 1) subjective experience (qualia); 2) conscious awareness and reflection, as best modeled by the theory of the "global workspace" in which the brain makes decisions; 3) self-consciousness, as being critically aware of, and reflecting about, one's own cognitive processes.

(1) is in my view much less mysterious than generally assumed, and relatively easy to implement in the Global Brain. For me, subjective experience is the implicit anticipation and evaluation that our brain makes based on (a) the presently incoming information (sensation, perception); (b) the associations we have learned through previous experience in which similar perceptions tended to co-occur with particular other phenomena (other perceptions, thoughts, emotions, evaluations). This creates an affectively colored, fuzzy pattern of expectation in which phenomena that are associated in this way are to some degree "primed" for possible use in further reflection or action.

Ben

So you're basically equating qualia with a certain informational process. But doesn't this ignore what Chalmers has called the "hard problem", i.e. the gap between subjective experience and physical reality? Or are you saying the qualia are associated with an *abstract process,* which is then instantiated in physical reality in a certain way? A little more clarification on your view toward the so-called "hard problem of consciousness" might be helpful for understanding your remarks.

Francis

I'd rather not get into that, or we will be gone for hours of discussion. I consider the "hard problem" as merely a badly formulated problem. Of course, things feel different from the inside and from the outside: I cannot feel what you can feel, but as long as you behave more or less similarly to how I might behave in similar circumstances, I will assume that you have feelings (qualia) similar to mine. It really does not matter whether you are "in reality" a human being, a zombie, a robot, or a GB: what counts is how you behave.

If you really want to go deeper into this, here are some of my recent writings in which I discuss the "hard problem":

1. Cognitive Systems: a cybernetic perspective on the new science of the mind[156]

2. Self-organization of complex, intelligent systems: An action ontology for transdisciplinary integration[157]

Ben

Yes, I see... Like our dear departed friend Valentin Turchin, you basically make the hard problem go away by assuming a monist ontology. The "action ontology" you describe is quite similar to things Val and I used to talk about (and I'm sure you guys evolved these ideas together, to some extent). You assume *action* as a primary entity, similarly to Whitehead with his process metaphysics (or Goethe, whose Faust said "In the Beginning was the Act!"), and then you think about states and

---

[156] http://pespmc1.vub.ac.be/Papers/CognitiveSystems.pdf

[157] http://pespmc1.vub.ac.be/Papers/ECCO-paradigm.pdf

objects and people and so forth ultimately as collections of actions.

This quote from your second link seems a critical one:

> *The ontology of action has no difficulty with subjective experience, and therefore it denies that there is an intrinsically "hard" problem of consciousness. First, it is not founded on the existence of independent, material objects obeying objective laws. Therefore, it has no need to reduce notions like purpose, meaning or experience to arrangements of such mechanical entities. Instead, it takes actions as it point of departure. An action, as we defined it, immediately entails the notions of awareness or sensation (since the agent producing the action needs to sense the situation to which it reacts), of meaning (because this sensation has a significance for the agent, namely as the condition that incites a specific action), and of purpose (because the action is implicitly directed towards a "goal", which is the attractor of the action dynamics).*

I can buy that, though I may interpret it a little differently than you do – to me it feels like a form of panpsychism, really. Mind is everywhere, matter is everywhere, and qualia are an aspect of everything.

Francis

Indeed, as I point out in that paper, panpsychism is a possible interpretation of my position. So is animism, the belief associated with "primitive" cultures, according to which all entities, including rocks, clouds and trees, are intentional agents. But I find such interpretations, while logically not incorrect, misleading, because they come with a baggage of irrational, mysterious, and semi-religious associations. The action ontology is really very simple, concrete and practical, and is intended for application in everyday life as well as in advanced agent-based technologies. In principle, it can even be formulated in mathematical form, as Valentin Turchin had started to do in his papers.

Ben

But what does this tell us about the qualia of the GB?

Francis

The process by which qualia emerge in the brain is not essentially different from the way collaborative filtering algorithms, after watching the choices we make (e.g. examining a number of books and music CDs on Amazon), produce an anticipation of which other items we might be interested in, and offer those as a recommendation. This is a purely subjective, fuzzy and constantly shifting list of potentially valuable options, which changes with every new choice we make. It may at this moment not "feel" anything like the qualia of our own concrete experiences, but that is mainly because these qualia reside inside our own brain, while the ones of the GB by definition are distributed over an amalgam of databases, hardware and people, so that no individual agent can claim to have direct access to them.

Ben

OK ... So then if we deal with qualia in this sort of way, we're still left with the problem of the theater of reflective consciousness – with the "global workspace" and with self-reflection.

Francis

The global workspace is based on the idea that difficult problems require full attention (i.e. maximal processing power) in which all specialized modules of the brain may need to be mobilized to attend to this particular problem. Reaching or interconnecting all modules at once requires a "global" (at the level of the individual brain, not at the planetary level) workspace through which such an important problem is "broadcasted", so that all modules can work on it. This implies a bottleneck, as only one problem can be broadcasted at a time in the human brain. This explains the sequential nature of conscious reflection, in contrast with the fact that subconscious processing in the brain is essentially parallel in nature.

At this time, I don't see any particular reason why the GB would develop such a bottleneck. Its processing resources (billions of people and their computers) are so huge that they can deal with many difficult problems in parallel.

Ben

Hmmm.... The current GB is not organized in such a way as to explicitly attack problems massively harder than those individual humans could attack. (Though it may implicitly attack them.) But I wonder if a future GB *could* explicitly try to solve problems significantly bigger and harder than those that any human can solve. These would then give rise to bottlenecks such as those you describe.

Francis

This is indeed an area worth of further investigation.

On the other hand, whether or not serious bottlenecks ever arise in GB information processing, the GB does seem to have use for some form of broadcasting: some problems may be so universal or urgent that ALL parts of the GB may need to be warned of it simultaneously. An example would be a terrorist attack of the scale of 9/11, an emerging pandemic, or contact made with an alien civilization.

In practice, the level of broadcasting will scale with the relative importance of the problem. A revolution in a Middle Eastern country, e.g., will catch the attention of most people in the Middle East, and of political and economic decision makers in most other parts of the world, but probably not of Latin American farmers. This selective broadcasting is what news media have been doing for decades, but their selection is rather biased by short-term political and economic motives. Hopefully, the emerging GB will do a better job of attending us to events and problems outside our immediate realm of interest. One example of how this may happen is how Google or other search engines select the "most important" websites or news items (as pointed out to me by Rome Viharo).

Ben

Right – but this kind of broadcasting seems fairly heterogeneous, rather than having a common hub like the global workspace, like the brain's executive networks, at the moment. But as the GB evolves and deals with more complex problems on the global level, it seems possible some sort of global workspace might arise. Related to this, an idea I had some time ago – and presented at the GB0 workshop – was to use an advanced AI system as basically an engineered global workspace for the GB.

But it's probably best not to diverge onto my AGI schemes and visions! So let's proceed with the aspects of consciousness and their manifestation in the GB. You've talked about qualia, broadcasting and the global workspace – what about self-reflection in the GB?

Francis

Certainly, self-reflection appears like a useful feature for the GB to have. Again, this does not seem to be so tricky to implement, as we, in our role of components of the GB, are at this very moment reflecting about how the GB functions and how this functioning could be improved... Moreover, decades ago already AI researchers have developed programs that exhibited a limited form of self-improvement by monitoring and manipulating their own processing mechanisms.

Ben

Any specific thoughts about how self-reflection might be implemented in the GB?

Francis

Not really, except that in an older paper[158] I sketched a simple methodology for "second-order learning", i.e. learning not only

---

[158] http://pespmc1.vub.ac.be/papers/Digital_libraries.pdf

the best values for associations between items, but the best values for the different parameters that underlie the learning algorithm, by comparing the predictions/recommendations made for different values of the parameters and seeing which fit best with reality/user satisfaction.

Another possible approach may be Valentin Turchin's approach of "metacompilation," a direct application of metasystem transitions to programming languages (which may be extendable to sufficiently powerful AI inference engines).

Ben

Metacompilation takes a computer program and represents its run-time behavior in a certain abstracted form, that lets it be very powerfully optimized. As you know I worked with Val and his colleagues a bit on the Java supercompiler, which was based on these principles. But to apply that sort of idea to the GB would seem to require some kind of very powerful "global brain metacompiler" oriented toward expressing the dynamics of aspects of the GB in an explicit form. Maybe something like what I was talking about before, of making a powerful AI to explicitly serve as the GB's global workspace.

But one thing that jumps out at me as we dig into these details, is how different the GB is from the human brain. It's composed largely of humans, yet it's a very very different kind of system. That brings up the question how you might compare the degree of intelligence of a global brain to that of a human? How smart is the Internet right now? How can one devise a measure of intelligence that would span different levels in this way; or do you think that's a hopeless intellectual quest?

Francis

I rather consider it a hopeless quest. Intelligence, like complexity, is at best represented mathematically as a *partial order*: for two random organisms A and B (say, a hedgehog and a magpie), A may be more intelligent than B, less intelligent, or equally intelligent, but most likely they are simply incomparable. A may be able to solve problems B cannot handle, but B can find solutions that A would not have any clue about.

For such a partial order, it is impossible to develop a quantitative measure such as an IQ, because numbers are by definition fully ordered: either IQ (A) < IQ (B), IQ (A)>IQ (B), or IQ (A)=IQ (B). IQ only works in people because people are pretty similar in the type of problems they can in principle solve, so by testing large groups of people with questions that do not demand specialized knowledge you can get a relatively reliable statistical estimate of where someone is situated with respect to the average (avg(IQ)=100), in terms of standard deviations (sigma(IQ)=15).

There is no average or standard deviation once you leave the boundaries of the human species, so there is no basis for us to evaluate the intelligence of something as alien as a Global Brain. At most, you might say that once it is fully realized, the GB will be (much, much) more intelligent than any single human.

Ben

In Shane Legg and Marcus Hutter's definition of Universal Intelligence[159], they define intelligence by basically taking a weighted average over all possible problems. So the intelligence of a creature is the average over all problems of its capability at solving that problem (roughly speaking; they give a rigorous mathematical definition). But of course, this means that intelligence is relative to the mathematical "measure" used to define the weights in the weighted average. So relative to one measure, a hedgehog might be more intelligent; but relative to another, a magpie might be more intelligent. In some cases system A might be better than system B at solving every possible problem, and in that case A would be smarter than B no matter what measure you choose.

---

[159] http://www.vetta.org/documents/UniversalIntelligence.pdf

Francis

This definition will not only be relative to the measure you choose, but also to the set of "all possible problems" to which you apply that measure. I do not see any objective way of establishing what exactly is in that set, since the more you know, the more questions (and therefore problems) you can conceive. Therefore, the content of that set will grow as your awareness of "possible problems" expands...

Ben

Well, from a mathematical point of view, one can just take the set of problems involved in the definition of intelligence to be the space of all computable problems – but indeed this sort of perspective comes to seem a bit remote from real-world intelligence.

But the practical key point is, you think the human brain and the GB right now are good at solving different kinds of problems, right? So in that case the assessment of which is more intelligent would depend on which problems you weight higher – and their intelligences aren't comparable in any objective sense.

OK, so if that's a hopeless quest, let's move on to something else – let's get a little more practical, perhaps. I'm curious, what technologies that exist right now do you think are pushing most effectively toward the creation/emergence of an advanced global brain?

Francis

I would mention three technologies that have been deployed extremely quickly and effectively in the last ten years:

1. Wikis (and related editable community websites) provide a very simple and intuitive medium for people to develop collective knowledge via the mechanism of stigmergy (activity performed by individuals leaving a trace on a shared site that incites others to add to that activity). Wikipedia is the most successful example: in ten years' time it developed

from nothing into the largest public knowledge repository ever conceived, which may soon contain the sum of all human knowledge.

2. Collaborative filtering or recommendation systems. This is the technology (based on closely guarded algorithms) used by sites such as YouTube and Amazon to recommend additional books, videos or other items on the basis of what you liked, and what others like you have liked previously. Unlike wikis, this is a collective intelligence technology that relies on *implicit* data, on information that was rarely consciously entered by any individual, but that can be derived relatively reliable from what that user did (such as ordering certain books, or watching certain videos rather than others). If wiki editing is similar to the rational, conscious reflection in the brain, collaborative filtering is similar to the subconscious, neural processes of selective strengthening of links and spreading activation.

3. Smartphones such as the iPhone, that make it possible to tap into the global brain at any time and any place. From simple person-to-person communication devices, these have morphed into universal, but still simple and intuitive interfaces that connect you to all the information that is globally available. This adds a very practical real-time dimension to GB problem-solving: when you need to get from A to B at time T, you want to know which means of transport you should take here and now; you are not interested in a full bus schedule. Thanks to in-built sensing technologies, such as a GPS, a compass, a camera and a microphone, a smart phone can first determine your local context (e.g. you are standing in front of the Opera building at sunset facing West while hearing some music playing in the background), then send that information to the GB together with any queries you may have (e.g. what is that melody? who designed that building? where can I get a pizza around here?), and finally relay the answer back to you.

Such ubiquitous access to the GB will not only help you to solve problems more quickly, but help the GB to gather more detailed and realistic data about what people do and what they need most (e.g. if many people wonder who designed that building, it may be worth installing a sign with the name of the architect, and if many people come to watch that building around sunset, it may be worth setting up bus lines that reach that destination just before sunset, and go back shortly afterwards).

Note that I more or less forecast the spread of technologies (2) and (3) in my first (1996) paper on the Global Brain, but somehow neglected the (in hindsight pretty obvious) contribution of (1). On the other hand, I forecast the spread of something more akin to Semantic Web supported, AI-type inference, but that has as yet still to make much of a splash...

Ben
Hmmm... So why do you think the Semantic Web hasn't flourished as you thought it would?

My own suspicion is that not many Web page authors were willing to mark up their web pages with meta-data, lacking any immediate practical reason to do so. Basically the semantic web is only useful if a large percentage of websites use it. The more websites use it, the more useful it is, and the more incentive there is for a new website to use it – but no incentive existed to get enough of a critical mass to use it, so as to start off an exponential growth process of increasing semantic web usage.

On the other hand, if we had sufficiently powerful AI to mark up Web pages automatically, thus making the Web "semantic" without requiring extra human effort, that would be a different story, and we'd have a different sort of semantic Web.

What's your take on the reasons?

Francis

I think you are partly right. Another reason is that semantic web people have generally underestimated the difficulty of building a consensual, formally structured ontology. The world tends to be much more contextual and fuzzy than the crisp categories used in logic or ontology.

Ontologies only work well in relatively restricted formal domains, such as names, addresses and telephone numbers. It already becomes much more difficult to create an ontology of professions, since new types of occupations are constantly emerging while old ones shift, merge or disappear. But if you stick to the formal domains, the semantic web approach does do not much more than a traditional database does, and therefore the added intelligence is limited.

I see the solution in some kind of a hybrid formal/contextual labeling of phenomena, where categories are to some degree fuzzy and able to adapt to changing contexts. An example of such a hybrid approach are user-added "tags", where the same item may get many different tags that are partly similar, partly overlapping, partly independent, and where tags get a weight simply by counting the number of people who have used a particular tag. But reasoning on tag clouds will demand a more flexible form of inference than the one used in semantic networks, and more discipline from the users to come up with truly informative tags.

Ben

And what sort of research are you working on these days? Anything global brain related, and if so in what sense?

Francis

I am presently working on three related topics:

1. the paleolithic, hunter-gatherer lifestyle as a model of what humans have evolved to live like, and thus a good starting

point if you want to understand how we can optimize our physical and mental health, strength and well-being

2. the concept of challenge as the fundamental driver of action and development in all agents, human as well as non-human

3. the problem of coordination in self-organization: how can a collective of initially autonomous agents learn to collaborate in the most productive way without any central supervisor telling them how to do it

The three topics are related in that they are all applications of what I call the "ontology of challenge and action", which sees the world as being constituted out of actions and their agents, and challenges as situations that elicit those actions. The life of a hunter-gatherer is essentially a sequence of (mostly unpredictable) challenges – mostly minor, sometimes major. In contrast, our modern civilized life has tried to maximally suppress or exclude uncontrolled challenges (such as accidents, germs, hot and cold temperatures, wild animals). Without these challenges, the various human subsystems that evolution has produced to deal with these challenges (e.g. the immune system, muscles, fast reflexes) remain weak and underdeveloped, leading to a host of diseases and mental problems.

The link with self-organization is that the action of one agent will in general change the environment in such a way as to produce a challenge to one or more other agents. If these agents react "appropriately", their interaction may become cooperative or synergetic; otherwise it is characterized by friction. In the best case, patterns of synergetic interaction propagate via the challenges they produce to the whole collective, which thus starts to act in a coordinated fashion.

This topic is obviously related to the Global Brain, which is such a self-organizing collective, but whose degree of coordination is obviously still far from optimal. I don't yet know precisely how, but I am sure that the notion of challenge will help me to better envision the technologies and requirements for such a collective coordination. One relevant concept I have called "mobilization system": A medium that stimulates people to act in a coordinated way by providing the right level of challenge. Again, Wikipedia is a prime example. The challenge here is: Can you improve in some way the page you have in front of you?

Ben

Hmmm...The notion of coordinating the GB reminds me of a broader issue regarding the degree of human coordination and planning and engineering required to bring about a maximally intelligent GB.

At the GB0 workshop in 2001, there seemed to be two major differences of opinion among participants on this (as well as very many smaller differences!). The first was whether the global brain was already present then (in 2001) in roughly the same sense it was going to be in the future; versus whether there was some major phase transition ahead, during which a global brain would emerge in a dramatically qualitatively stronger sense. The second was whether the emergence of the global brain was essentially something that was going to occur "spontaneously" via general technological development and social activity; versus the global brain being something that some group of people would specifically *engineer* (on top of a lot of pre-existing technological and social phenomena). Of course I've just drawn these dichotomies somewhat crudely, but I guess you understand the ideas I'm getting at. What's your view on these dichotomies and the issues underlying them?

Francis

My position is nicely in the middle: either position on each of the dichotomies seems too strong, too reductionistic to me. I believe that the GB to some degree is already there in essence, to some degree it still reserves a couple of spectacular surprises for us over the coming decades. Similarly, it will to some degree emerge spontaneously from the activities of many, relatively clueless people, to some degree be precipitated by clever engineering, inspired by the ideas of visionary thinkers such as you or I!

Ben

Another, related question is the connection between the GB and the Singularity. I take it you're familiar with Ray Kurzweil's and Vernor Vinge's notion of the Singularity. What's your current take on this notion? Is the Singularity near? As I vaguely recall, when we discussed this once before you were a little skeptical (but please correct me if I'm wrong). Max More likes to talk about a Surge rather than a Singularity – a steady ongoing growth of advanced technology, but without necessarily there being any point of extremely sudden and shocking advance. His Surge would ultimately get us to the same (radically transhuman) point as Kurzweil's Singularity, but according to a different slope of progress. Are you perhaps more friendly to Max's Surge notion than Ray and Vernor's Singularity? Or do you find them both unjustifiably techno-optimistic?

Francis

I have just been invited to write a paper for a special volume that takes a critical look at the Singularity. I do not know what exactly Max More means by his Surge, but it does sound more realistic than a true Singularity. In the paper, I plan to argue that the transition to the Global Brain regime is more likely to resemble a logistic or S-curve, which starts to grow nearly exponentially, then slows down to a near linear expansion (constant growth), in order to finally come to a provisional halt (no more growth).

In my own (decidedly subjective experience), we may already be in the phase of constant growth, as I have the feeling that since about the year 2000 individuals and society are to such a degree overwhelmed with the on-going changes that their creativity and capacity for adaptation suffers, thus effectively slowing down further innovation. This doesn't mean that we should no longer expect spectacular innovations, only that they will no longer come at an ever increasing speed.

That may seem defeatist to Singularitarians and other transhumanist enthusiasts, but I believe the basic infrastructure, technical as well as conceptual, for the Global Brain is already in place, and just needs to be further deployed, streamlined and optimized. We have only glimpsed a mere fraction of what the GB is capable of, but realizing its further potential may require fewer revolutionary innovations than one might think.

Ben

Yes, I see… You recently wrote on a mailing list that

> In summary, my position is
>
> 1) I believe in the Singularity as a near-term transition to a radically higher level of intelligence and technological power, best conceived of as a "Global Brain."
>
> 2) I don't believe in the Singularity as a near-term emergence of super-intelligent, autonomous, computers.
>
> 3) I don't believe in the Singularity as a near-term acceleration towards a practically infinite speed of technological and economic progress.

I think that puts it pretty clearly.

And as you know, I don't fully agree. I agree that a Global Brain is emerging, but I see this as part of the dawning Vingean Singularity, not as an alternative. I think superintelligent computers will emerge and that eventually they will be able to operate quite autonomously of humans and human society – though initially our first superintelligent computers will probably be richly enmeshed with the Global Brain. And I do think we'll have acceleration toward an incredibly rapid speed of technological and economic progress – though I also think that, from the perspective of human society, there's a limit to how fast things can progress, because there's a limit to how fast human beings can absorb and propagate change. There's also probably a limit to how MUCH things can change for humans, given the constraint of humans remaining humans. The way I see it, at some point future history is likely to bifurcate – on the one hand you'll have advanced AIs integrated with humans and the Global Brain, advancing at an impressive but relatively modest pace due to their linkage with humans; and on the other hand you'll have advanced AIs detached from the human context, advancing at a pace and in a direction incomprehensible to legacy humans. Some people fear that if AIs advance in directions divergent from humanity, and beyond human ken, this will lead to the destruction of humankind; but I don't see any clear reason why this would have to be the case.

In practical terms, I think that in the next few decades (possibly sooner!), someone (maybe my colleagues and me) is going to create human-level (and then transhuman) artificial general intelligence residing in a relatively modest-sized network of computers (connected into and leveraging the overall Internet as a background resource). Then, I think the integration of this sort of AGI into the GB is going to fundamentally change its character, and drastically increase its intelligence. And then after that, I think some AGIs will leave the Global Brain and the whole human domain behind, having used humanity and the GB as a platform to get their process of self-improvement and learning started.

I'm particularly curious for your reaction to this possibility.

Francis

My personal bias is to consider the "background resource" of knowledge available on the Internet more important than the localized AGI. Such an AGI would definitely be very useful and illuminating to have, but without the trillions of (mostly) human generated data available via the net, it wouldn't be able to solve many real-life problems. This perspective comes from the situated and embodied cognition critique on AI (and by extension AGI): Real intelligence only emerges in constant interaction with a truly complex and dynamic environment. The higher the bandwidth of that interaction, the more problems can be solved, and the more pragmatically meaningful the conclusions reached by your intelligent system become.

The only practical way I see at the moment to maximize that bandwidth is to use all the globally available sensors and effectors, i.e. all human individuals supported by their smartphone interfaces, plus a variety of autonomous sensors/effectors built into the environment, as envisaged by the "ambient intelligence/ubiquitous Internet" paradigm. That means in effect that your intelligent system should be firmly rooted into the GB, extending its "feelers" into all its branches and components.

Whether your AGI system runs locally on a modest size network of computers, or in distributed form on the Internet as a whole seems rather irrelevant to me: this is merely a question of hardware implementation. After all, nobody really cares where Google runs its computers: What counts are the way they sieve through the data...

By the way, when discussing these issues with my colleague Mark Martin, he mentioned Watson, an IBM system perhaps not unlike what you envisage. While the IBM website is extremely vague about how Watson is supposed to answer the questions posed to it, I suspect that it too is firmly rooted into an Internet-scale database of facts, texts and observations gathered by millions of people.

Of course, IBM has reason to downplay the role of those (publically available) data, and to emphasize the great strides they made on the level of hardware (and to a lesser degree) software, just like you would rather focus on the advanced AGI architecture underlying your system. But my impression is that neither system would be of much practical use without that humongous database of human-collected information behind it.

Ben

Well, as you know Watson is not an artificial general intelligence – Watson is just a fairly simple question-answering system, that responds to questions based on looking up information that's already present on the Web in textual form. So, for sure, in the case of a system like Watson, the AI algorithms play a secondary role to the background knowledge. But that's because Watson is not based on a serious cognitive architecture that tries to learn, to self-reflect, to create, to model the world and its place in the world and its relationship to others.

Systems like Watson are relatively easy to build and fun to play with, precisely because they're just tools for leveraging the knowledge on the Web. But they're a completely different animal from the kind of AGI system we're trying to build in the OpenCog project, for example (or from the human brain, which also is capable of wide-ranging learning and creativity, not just matching questions against a database of previously-articulated answers).

The knowledge available on the Web will also be of tremendous use to real AGI systems – but unlike Watson these systems will do something besides just extract knowledge from the Web and respond with it appropriately. They will do more like humans do – feed the knowledge from the Web into their own internal thought processes, potentially creating ideas radically different from anything they read. Like you or me, they will question everything they read and even whether 1+1=2. What happens when systems like this become very powerful and intelligent and interact intensively with the GB is an interesting question. My view is that, at a certain point, AGI minds will come to dominate the GB's dynamics (due to the AGIs eventually becoming more generally intelligent than humans); and that the GB will in essence serve as an incubator for AGI minds that will ultimately outgrow the human portion of the GB.

But, I know you don't share that particular portion of my vision of the future of the GB – and I'd be the first to admit that none of us knows for sure what's going to happen.

Steve Omohundro:
The Wisdom of the Global Brain and the Future of AGI

Interview first published April 2011

The future of humanity involves a complex combination of technological, psychological and social factors. One of the difficulties we face in comprehending and crafting this future is that not many people or organizations are adept at handling all these aspects. Dr. Stephen Omohundro's approach is one of the fortunate exceptions to this general pattern, and this is part of what gives his contributions to the futurist domain such a unique and refreshing twist.

Steve has a substantial pedigree and experience in the hard sciences, beginning with degrees in Mathematics and Physics from Stanford and a Ph.D. in Physics from U.C. Berkeley. He was a professor in the Computer Science department at the University of Illinois at Champaign-Urbana. Co-founded the Center for Complex Systems Research, authored the book "Geometric Perturbation Theory in Physics", and designed the programming languages StarLisp and Sather. He wrote the 3D graphics system for Mathematica, and built systems which learn to read lips, control robots, and induce grammars. I've had some long and deep discussions with Steve about advanced artificial intelligence, comparing my own approach to his own unique AI designs.

Steve has also developed considerable expertise and experience in understanding and advising human minds and systems. Via his firm Self-Aware Systems[160], he has worked with clients using a variety of individual and organizational change processes including: Rosenberg's Non-Violent Communication,

---

[160] http://selfawaresystems.com/

Gendlin's Focusing, Travell's Trigger Point Therapy, Bohm's Dialogue, Beck's Life Coaching, and Schwarz's Internal Family Systems Therapy.

Steve's papers and talks on the future of AI, society and technology include: *The Wisdom of the Global Brain*[161] and *Basic AI Drives* [162]. Both reflect his dual expertise in technological and human systems.

In this interview, I was keen to mine his insights regarding both the particular issue of the risks facing the human race as we move forward along the path of accelerating technological development.

Ben

A host of individuals and organizations – Nick Bostrom, Bill Joy, the Lifeboat Foundation, the Singularity Institute, and the Millennium Project, to name just a few – have recently been raising the issue of the "existential risks" that advanced technologies may pose to the human race. I know you've thought about this topic a fair bit yourself, both from the standpoint of your own AI work, and, more broadly. Could you share the outlines of your thinking in this regard?

Steve

I don't like the phrase "existential risk" for several reasons. It presupposes that we are clear about exactly what "existence" we are risking. Today, we have a clear understanding of what it means for an animal to die or a species to go extinct. However, as new technologies allow us to change our genomes and our physical structures, it will become much less clear to us when we lose something precious. "Death" and "extinction," for example,

---

[161] http://selfawaresystems.com/2009/12/24/the-wisdom-of-the-global-brain/

[162] http://selfawaresystems.com/2007/11/30/paper-on-the-basic-ai-drives/

become much more amorphous concepts in the presence of extensive self-modification.

It's easy to identify our humanity with our individual physical form and our egoic minds, but in reality, our physical form is an ecosystem, only 10% of our cells are "human." Our minds are also ecosystems composed of interacting subpersonalities. Our humanity is as much in our relationships, interconnections, and culture as it is in our individual minds and bodies. The higher levels of organization are much more amorphous and changeable. For these reasons, it could be hard to pin down what we are losing at the moment when something precious is lost. {Addition to clear with Steve: It is more likely that we will only realize what we have lost, long after it's already gone for good.}

So, I believe the biggest "existential risk" is related to identifying the qualities that are most important to humanity and to ensuring that technological forces enhance those rather than eliminate them. Already today we see many instances where economic forces act to create "soulless" institutions that tend to commodify the human spirit rather than inspire and exalt it.

Some qualities that I see as precious and essentially human include: love, cooperation, humor, music, poetry, joy, sexuality, caring, art, creativity, curiosity, love of learning, story, friendship, family, children, etc. I am hopeful that our powerful new technologies will enhance these qualities. But I also worry that attempts to precisely quantify them may in fact destroy them. For example, the attempts to quantify performance in our schools using standardized testing have tended to inhibit our natural creativity and love of learning.

Perhaps the greatest challenge that will arise from new technologies will be to really understand ourselves and identify our deepest and most precious values.

Ben

Yes... After all, "humanity" is a moving target, and today's humanity is not the same as the humanity of 500 or 5000 years ago, and humanity of 100 or 5000 years from now – assuming it continues to exist – will doubtless be something dramatically different. But still there's been a certain continuity throughout all these changes, and part of that doubtless is associated with the "fundamental human values" that you're talking about.

Still, though, there's something that nags at me here. One could argue that none of these precious human qualities are practically definable in any abstract way, but they only have meaning in the context of the totality of human mind and culture. So that if we create a fundamentally nonhuman AGI that satisfies some abstracted notion of human "family" or "poetry", it won't really satisfy the essence of "family" or "poetry". Because the most important meaning of a human value doesn't lie in some abstract characterization of it but rather in the relation of that value to the total pattern of humanity. In this case, the extent to which a fundamentally nonhuman AGI or cyborg or posthuman or whatever would truly demonstrate human values, would be sorely limited. I'm honestly not sure what I think about this train of thought. I wonder what's your reaction.

Steve

That's a very interesting perspective! In fact it meshes well with a perspective I've been slowly coming to, which is to think of the totality of humanity and human culture as a kind of "global mind". As you say, many of our individual values really only have meaning in the context of this greater whole. And perhaps it is this greater whole that we should be seeking to preserve and enhance. Each individual human lives only for a short time but the whole of humanity has a persistence and evolution beyond any individual. Perhaps our goal should be to create AGIs that integrate, preserve, and extend the "global human mind" rather than trying solely to mimic individual human minds and individual human values.

Steve Omohundro:
The Wisdom of the Global Brain and the Future of AGI

Ben

Perhaps a good way to work toward this is to teach our nonhuman or posthuman descendants human values by example, and by embedding them in human culture so they absorb human values implicitly, like humans do. In this case we don't need to "quantify" or isolate our values to pass them along to these other sorts of minds.

Steve

That sounds like a good idea. In each generation, the whole of human culture has had to pass through a new set of minds. It is therefore well adapted to being learned. Aspects which are not easily learnable are quickly eliminated. I'm fascinated by the process by which each human child must absorb the existing culture, discover his own values, and then find his own way to contribute. Philosophy and moral codes are attempts to codify and abstract the learnings from this process but I think they are no substitute for living the experiential journey. AGIs which progress in this way may be much more organically integrated with human society and human nature. One challenging issue, though, is likely to be the mismatch of timescales. AGIs will probably rapidly increase in speed and keeping their evolution fully integrated with human society may become a challenge.

Ben

Yes, it's been amazing to watch that learning process with my own 3 kids, as they grow up.

It's great to see that you and I seem to have a fair bit of common understanding on these matters. This reminds me, though, that a lot of people see these things very, very differently. Which leads me to my next question: What do you think are the biggest *misconceptions* afoot, where existential risk is concerned?

Steve

I don't think the currently fashionable fears like global warming, ecosystem destruction, peak oil, etc. will turn out to be the most important issues. We can already see how emerging technologies could, in principle, deal with many of those problems. Much more challenging are the core issues of identity, which the general public hasn't really even begun to consider. Current debates about stem cells, abortion, cloning, etc. are tiny precursors of the deeper issues we will need to explore. And we don't really yet have a system for public discourse or decision making that is up to the task.

Ben
Certainly a good point about public discourse and decision making systems. The stupidity of most YouTube comments, and the politicized (in multiple senses) nature of the Wikipedia process, makes clear that online discourse and decision-making both need a lot of work. And that's not even getting into the truly frightening tendency of the political system to reduce complex issues to oversimplified caricatures.

Given the difficulty we as a society currently have in talking about, or making policies about, things as relatively straightforward as health care reform or marijuana legalization or gun control, it's hard to see how our society could coherently deal with issues related to, say, human-level AGI or genetic engineering of novel intelligent lifeforms!

For instance, the general public's thinking about AGI seems heavily conditioned by science-fiction movies like Terminator 2, which clouds consideration of the deep and in some ways difficult issues that you see when you understand the technology a little better. And we lack the systems needed to easily draw the general public into meaningful dialogues on these matters with the knowledgeable scientists and engineers.

So what's the solution? Do you have any thoughts on what kind of system might work better?

Steve
I think Wikipedia has had an enormous positive influence on the level of discourse in various areas. It's no longer acceptable to

plead ignorance of basic facts in a discussion. Other participants will just point to a Wikipedia entry. And the rise of intelligent bloggers with expertise in specific areas is also having an amazing impact. One example I've been following closely are debates and discussions about various approaches to diet and nutrition.

A few years back, T. Colin Campbell's "The China Study[163]" was promoted as the most comprehensive study of nutrition, health, and diet ever conducted. The book and the study had a huge influence on people's thinking about health and diet. A few months ago, 22 year old English major Denise Minger decided to reanalyze the data in the study and found that they did not support the original conclusions. She wrote about her discoveries on her blog[164] and sparked an enormous discussion all over the health and diet blogosphere that dramatically shifted many people's opinions. The full story can be heard in her interview[165]

It would have been impossible for her to have had that kind of impact just a few years ago. The rapidity with which incorrect ideas can be corrected and the ease with which many people can contribute to new understanding is just phenomenal. I expect that systems to formalize and enhance that kind of group thinking and inquiry will be created to make it even more productive.

---

[163] http://www.amazon.com/China-Study-Comprehensive-Nutrition-Implications/dp/1932100660/ref=sr_1_1?ie=UTF8&qid=1297888115&sr=8-1

[164] http://rawfoodsos.com/2010/07/07/the-china-study-fact-or-fallac/

[165] http://livinlavidalowcarb.com/blog/livin-la-vida-low-carb-show-episode-405-denise-minger-exposes-some-major-flaws-in-the-china-study/9076

Ben

Yes, I see – that's a powerful example. The emerging Global Brain is gradually providing us the tools needed to communicate and collectively think about all the changes that are happening around and within us. But it's not clear if the communication mechanisms are evolving fast enough to keep up with the changes we need to discuss and collectively digest.

On the theme of rapid changes, let me now ask you something a little different – about AGI.... I'm going to outline two somewhat caricaturish views on the topic and then probe your reaction to them!

First of all, one view on the future of AI and the Singularity is that there is an irreducible uncertainty attached to the creation of dramatically greater than human intelligence. That is, in this view, there probably isn't really any way to eliminate or drastically mitigate the existential risk involved in creating superhuman AGI. So, in this view, building superhuman AI is essentially plunging into the Great Unknown and swallowing the risk because of the potential reward.

On the other hand, an alternative view is that if we engineer and/or educate our AGI systems correctly, we can drastically mitigate the existential risk associated with superhuman AGI, and create a superhuman AGI that's highly unlikely to pose an existential risk to humanity.

What are your thoughts on these two perspectives?

Steve

I think that, at this point, we have tremendous leverage in choosing how we build the first intelligent machines and in choosing the social environment that they operate in. We can choose the goals of those early systems and those choices are likely to have a huge effect on the longer-term outcomes. I believe it is analogous to choosing the constitution for a country. We have seen that the choice of governing rules has an enormous effect on the quality of life and the economic productivity of a population.

Ben

That's an interesting analogy. And an interesting twist on the analogy may be the observation that to have an effectively working socioeconomic system, you need both good governing rules, and a culture oriented to interpreting and implementing the rules sensibly. In some countries (e.g. China comes to mind, and the former Soviet Union) the rules as laid out formally are very, very different from what actually happens. The reason I mention this is: I suspect that in practice, no matter how good the "rules" underlying an AGI system are, if the AGI is embedded in a problematic culture, then there's a big risk for something to go awry. The quality of any set of rules supplied to guide an AGI is going to be highly dependent on the social context.

Steve

Yes, I totally agree! The real rules are a combination of any explicit rules written in law books and the implicit rules in the social context. Which highlights again the importance for AGIs to integrate smoothly into the social context.

Ben

One might argue that we should first fix some of the problems of our cultural psychology, *before* creating an AGI and supplying it with a reasonable ethical mindset and embedding it in our culture. Because otherwise the "embedding in our culture" part could end up unintentionally turning the AGI to the dark side!! Or on the other hand, maybe AGI could be initially implemented and deployed in such a way as to help us get over our communal psychological issues... Any thoughts on this?

Steve

Agreed! Perhaps the best outcome would be technologies that first help us solve our communal psychological issues and then as they get smarter evolve with us in an integrated fashion.

Ben

On the other hand, it's not obvious to me that we'll be able to proceed that way, because of the probability – in my view at any rate – that we're going to need to rely on advanced AGI systems to protect us from other *technological* risks.

For instance, one approach that's been suggested, in order to mitigate existential risks, is to create a sort of highly intelligent "AGI Nanny" or "Singularity Steward". This would be a roughly human-level AGI system without capability for dramatic self-modification, and with strong surveillance powers, given the task of watching everything that humans do and trying to ensure that nothing extraordinarily dangerous happens. One could envision this as a quasi-permanent situation, or else as a temporary fix to be put into place while more research is done regarding how to launch a Singularity safely.

Any thoughts on the sort of AI Nanny scenario?

Steve

I think it's clear that we will need a kind of "global immune system" to deal with inadvertent or intentional harm arising from powerful new technologies like biotechnology and nanotechnology. The challenge is to make protective systems powerful enough for safety but not so powerful that they themselves become a problem. I believe that advances in formal verification will enable us to produce systems with provable properties of this type. But I don't believe this kind of system on its own will be sufficient to deal with the deeper issues of preserving the human spirit.

Ben

What about the "one AGI versus many" issue? One proposal that's been suggested, to mitigate the potential existential risk of human-level or superhuman AGIs, is to create a community of AGIs and have them interact with each other, comprising a society with its own policing mechanisms and social norms and so forth. The different AGIs would then keep each other in line. A "social safety net" so to speak.

Steve

I'm much more drawn to "ecosystem" approaches which involve many systems of different types interacting with one another in such a way that each acts to preserve the values we care about. I think that alternative singleton "dictatorship" approaches could also work but they feel much more fragile to me in that design mistakes might become rapidly irreversible. One approach to limiting the power of individuals in an ecosystem is to limit the amount of matter and free energy they may use while allowing them freedom within those bounds. A challenge to that kind of constraint is the formation of coalitions of small agents that act together to overthrow the overall structure. But if we build agents that *want* to cooperate in a defined social structure, then I believe the system can be much more stable. I think we need much more research into the space of possible social organizations and their game theoretic consequences.

Ben

Finally – bringing the dialogue back to the practical and near-term – I wonder what you think society could be doing now to better militate against existential risks from AGI or from other sources?

Steve

Much more study of social systems and their properties, better systems for public discourse and decision making, deeper inquiry into human values, improvements in formal verification of properties in computational systems.

Ben

That's certainly sobering to consider, given the minimal amount of societal resources currently allocated to such things, as opposed to for example the creation of weapons systems, better laptop screens or chocolaty-er chocolates!

To sum up, it seems one key element of your perspective is the importance of deeper collective (and individual) self-understanding – deeper intuitive and intellectual understanding of the essence of humanity. What is humanity, that it might be preserved as technology advances and wreaks its transformative impacts? And another key element is your view is that social networks of advanced AGIs are more likely to help humanity grow and preserve its core values, than isolated AGI systems. And then there's your focus on the wisdom of the global brain. And clearly there are multiple connections between these elements, for instance a focus on the way ethical, aesthetic, intellectual and other values emerge from social interactions between minds. It's a lot to think about... But fortunately none of us has to figure it out on our own!

Alexandra Elbakyan: Beyond the Borg

Interview first published February 2011

The goal of Brain Computer Interfacing technology is simple to describe – connecting the biocomputers in our skulls with the silicon chip based computers on our desks and in our pockets. And it's also simple to imagine the vast array of life-changing possibilities this technology might provide once it matures. This interview highlights one of the most exciting of these potentials – the emergence of distributed mind networks that rely on direct brain-to-brain links.

I discussed the concept of using BCI to connect brains together a bit at the end of a 2009 H+ magazine article that I wrote, titled "Brain-Computer Interfacing: From Prosthetic Limbs to Telepathy Chips[166]". And the following year, at the 2010 *Toward a Science of Consciousness* conference in Tuscon, I was very pleased to encounter a young researcher – Alexandra Elbakyan – advocating a similar theme. Her poster[167] at that conference was mainly on merging human and machine consciousness using BCI, but in discussion she quickly turned to the potential to use BCI to create shared consciousness among multiple humans and machines. At the H+ Summit @ Harvard[168] conference later that year, she gave a talk[169] specifically on the latter theme: "Brain-Computer Interfacing, Consciousness, and the Global Brain: Towards the Technological Enlightenment"

---

[166] http://hplusmagazine.com/2009/07/13/brain-computer-interfacing-prosthetic-limbs-telepathy-chips

[167] http://engineuring.files.wordpress.com/2010/04/tsc2010_poster.png

[168] http://hplussummit.com/2010/east/

[169] http://engineuring.wordpress.com/2010/06/14/mind-implants-or-how-to-expand-consciousness-using-new-technologies/

Alexandra began her career with a computer science degree from Kazakh National Technical University in her home country of Kazakhstan, in addition to her coursework pursuing a research project on security systems that recognize individuals via their brain waves. After obtaining her degree she worked for a while with the Human Media Interaction Group at the University of Twente on the mind-controlled game "Bacteria Hunt[170]"; and then at the Human Higher Nervous Activity Lab at the Russian Academy of Sciences studying consciousness. Most recently she has been involved with the Brain Machine Interfacing Initiative at Albert Ludwigs University in Freiburg developing prosthetic hands, and has worked at Steve Potter's neuroengineering group at Georgia Tech. With all this diverse hands-on BCI experience plus a unique and broad long-term vision of the future potential of the technology, I was sure Alexandra would be a fascinating interview subject.

Ben

In your talk at the H+ Summit at Harvard University last June (2010), you said the following provocative words: "Potentially, brain chips can also be designed to have consciousness inside them. Inserted into a human brain, such a conscious implant would expand the user's conscious experience with its own contents. For this, however, a new kind of brain-machine interface should be developed that would merge consciousness in two separate systems – the chip and the brain – into single, unified one".

OK – this is pretty interesting and exciting-sounding! – but I have a few questions about it.

First – just to get this out of the way – what do you mean by "consciousness" when you use that word? As you know, different researchers give the word very different meanings.

---
[170] http://www.citeulike.org/user/cmuehl/article/7706119

Alexandra

Yep, the word "consciousness" spans a broad spectrum of meanings. I refer to the most common one, known as "phenomenal consciousness", or "qualia" in philosophers' language. Consciousness, in this sense, is hard to define or explain but very easy to understand. Simply put, it is the ability to experience the world around us – in pictures, sounds, smells – plus the stuff that is happening inside our brain and body – thoughts, memories, emotions and etc.

May I describe an experiment to illustrate this better: If I put you in front of a computer screen that will show some image for a very very short time (around 15 ms), you won't be able to see anything. But still, you will be able to learn and take actions based on this image. For example, if I give you a series of puzzle-solving tasks – some solvable and some not – and then flicker my invisible image for each unsolvable puzzle – then after, you'll be more likely to drop the task when exactly this image is flickered, regardless of whether the task is solvable or not. So you actually perceived the image, memorized it and made a decision based on it. And all this happened without you actually noticing anything! This effect is called "unconscious priming."

Such experiments show something exists that is different from mere thinking, perceiving, learning, acting, and living – something that enables us to actually *experience* what is going on – i.e., consciousness.

Ben

So what do you think are the ingredients that would enable a brain chip to have qualia (consciousness) inside it? Am I correct in assuming that your hypothesized brain chip is an ordinary digital computer chip? So that the consciousness inside the chip – the qualia – are achieved by appropriate programming?

Alexandra

I doubt the ordinary digital computer chips would be able to perform on par with the rest of the brain, given that we need a supercomputer to emulate even a single cortical column like in the Blue Brain[171] project. So it seems that ordinary digital chips wouldn't integrate well into real-time conscious experience.

It is more likely that a new class of information-processing hardware – neuromorphic chips[172] – that is currently under active development, will be useful for the project. These neuromorphic chips try to mimic approaches used by biological nervous systems. Probably, they'll be effective for recreating consciousness, too.

Ben

Do you feel like you have a good subjective sense of what it would feel like to merge your brain-consciousness with a chip-consciousness in that way? Would it feel like expanding yourself to include something new? Or could it maybe feel like losing yourself and becoming something quite different?

Alexandra

It depends on the implementation of the chip!

For example, we can (I hope) develop a chip that would "contain" visual qualia. Being connected to the brain, such chip will feel like a third eye, expanding our visual field to include pictures from the connected camera. This camera can be attached to one's back – providing a 360-degree panoramic worldview – or it may be connected to the Internet and project to the consciousness pictures from remote places. And implants with visual qualia will surely be helpful for blind persons, too.

But visual consciousness is by no means the only modality of consciousness: there are also sounds, smells, thoughts, and so on, and every modality is unique, very different from all others. Probably, within a chip we can even engineer absolutely new

---

[171] http://bluebrain.epfl.ch/
[172] http://www.neurdon.com/2010/08/12/neuromorphic-systems-silicon-neurons-and-neural-arrays-for-emulating-the-nervous-system/

kinds of conscious modalities. But how is it possible to imagine what these modalities will feel like?

Ben

Hmmm... But taking the visual-qualia chip as an example: Would this chip be conscious in the sense that the brain is conscious, or merely conscious in the sense that the *eye* is conscious?

Alexandra

Neither! The chip would be conscious in the sense that any *part* of the brain is conscious (i.e. any one of the parts that is directly involved in conscious experience, from which the retina is excluded so far as I know)

Ben

A more interesting case would be a chip that would extend the human short-term memory capability beyond 7 +/-2[173] to say 50 +/- 10... Because this STM chip would be inserted right at the heart of our everyday conscious experience, our "theater of conscious experience" or "global workspace." In this case, there seem to be two possibilities.

The first possibility is that our qualia, our subjective experience, would really feel extended over all 50 items in the STM.

The second possibility is that the regular old STM with 7+/- 2 capacity would feel like "our consciousness" and the other 43 or so new "memory slots" would just be feeding data into our awareness, as if they were an external data source.

Which possibility are you foreseeing would happen when the STM brain chip was implanted? I'm getting the feeling that you're expecting the first one.

---

[173]
http://en.wikipedia.org/wiki/The_Magical_Number_Seven,_Plus_or_Minus_Two

Alexandra

Hmmm... Both possibilities are interesting, actually.

But please remember that in order to think on the information stored in working memory, we need to bring it into awareness anyway!

I agree that our brains are capable of sophisticated unconscious problem-solving too, but in most cases, to think clearly, we need to be aware of what we're thinking about. Probably, that's why sketchpads, mindmaps and so forth are so useful for brainstorming – because we can simultaneously bring a collection of essential points into awareness by simply looking at the paper or screen.

In other words, expanding working memory without increasing awareness of its contents, will simply enable us to quickly remember more details. By expanding capabilities of our consciousness, we'll be able to easily process these details simultaneously while thinking.

Ben

So... You think both of my possibilities are feasible, but the first possibility is a lot more interesting, because a working memory that's largely unconscious wouldn't be capable of such advanced cognition as one that was more fully within consciousness.

Alexandra

Yes - in addition to the fact that being conscious of what is going on in one's brain – especially when it is busy solving some problem – is simply more interesting than just acting like a zombie, consciousness seem to play an important role in thinking itself.

Ben

OK, now I have a more general sort of question. I'm wondering about the relation between "Ben with a 50-item working memory capacity" and "the present, stupid old legacy Ben with a mere

7+/-2 working memory capacity." Would this new mentally-super Ben still feel like me?

In other words: I wonder how much our self-system is dependent on our *limitations*? And when these limitations are lifted, maybe in some sense we aren't really ourselves anymore?

Alexandra

But we lift our limitations by learning skills and acquiring knowledge and progress from ourselves to non-ourselves every day. I'm not at all the same person as I was 5 years ago, and so on...

Ben

That's true, but some changes are bigger and more sudden than others. I've often wondered if some level of continuity of change is necessary to preserve the feeling of self. Since self is in some ways more a process than a state.

So maybe I should increase my STM capacity by one item each week, instead of going from 7 to 50 all at once!

Speaking of big changes to the mind, though -- you've also discussed the possibility of using brain-computer interfaces to join multiple people together into a kind of global consciousness. This also gives rise to a lot of questions (an understatement, I guess!).

For instance, do you think it would be possible to feed one person's thoughts somehow directly into another person's brain, via interfacing the chips installed in their respective brains? Or do you think some kind of translation or normalization layer would be required between the two people, to translate one person's individual thought-language into the other person's individual thought-language?

Alexandra

Actually, "feeding" thoughts from one brain to another, or even "translating" them is the complete opposite of the shared-consciousness concept!

Within shared consciousness we'll be able to perceive thoughts that are going on in other brains directly, so there is no need to copy them anywhere. To explain, brain activity gives rise to unified conscious experience; but still, for this to happen, each separate neuron does not have to receive a copy of all thoughts that are being thought by all other neurons – instead, each neuron simply contributes its own part to the whole experience. The same holds when connecting minds of different people together: To bring up an experience of shared consciousness, one doesn't require explicit transfer of thoughts from one brain to another.

But what information, if not thoughts, will be transferred through the connection then? That's one of the things we need to figure out!

Ben

Hmmm... I wonder if some of my own work on cognitive modeling may give some clues about that. One can model memory as "glocal", i.e. memories can be modeled as containing localized aspects, but also global distributed aspects. The localized aspect of a memory can then trigger the activation of the global aspect, in the overall distributed mind network. This is how things work in our OpenCog[174] AI system. So if thoughts have localized and globalized aspects, maybe it's the localized aspects that get passed around, causing the globalized aspects to self-organize in the minds that receive the messages.

But, getting back to the experiential aspect – do you feel like you have a good subjective sense of what it would feel like to mentally link with other peoples' minds in this way? Would it feel

---

[174] http://opencog.org/

like becoming part of a greater whole, an uber-mind or more like having other people's thoughts flit in and out of one's head? Or might the character of the experience depend sensitively on the details of the technical implementation?

Alexandra

I thought about what the collective consciousness can feel like, and I can imagine three options here

First of all, there's "Borg-like" consciousness. This will feel like our experience, knowledge, and worldview have been tremendously expanded. We will have access to one another's memories, feelings, and expertise, and therefore will be able to think more broadly, make better decisions, and generate better ideas.

Then there's what I call "Background" consciousness. In this option, we will experience several simultaneous streams of consciousness, but these streams won't influence each other in any way. We will be aware of several contradictory viewpoints (coming from different brains) at the same time, but still won't be able to make decisions or generate ideas that take into account each of the viewpoints (unless connected people will discuss with each other, of course)

Finally – and most excitingly – there's "Emergent" or "higher-level" consciousness. This one comes out of the idea that a group of people may itself constitute some kind of an organism. Just like biological cells, collected together, constitute a human being with higher levels of intelligence and consciousness than possessed by any individual cell, similarly a group of people can constitute yet another kind of being, that is capable of its own thinking and experience.

Actually, recent research [175] has shown that collective intelligence within a group of people is actually present and irreducible to the intelligence of individual group members. This IQ arises from interactions between them. It is very possible that collective consciousness arises when people are brought together, too; and we may potentially experience this higher level of consciousness by physically connecting our brains through BMI (brain-machine interfacing).

Ben
Interesting possibilities! But, I'm afraid I don't fully understand your distinction between Borg consciousness and emergent consciousness – could you clarify?

Alexandra
Emergent consciousness is different from Borg consciousness because in the Borg case, the content of consciousness is expanded. We have more ideas, and more sources that generate ideas, more experience, and so on, inside our mind; but still experience is basically the same.

Emergent consciousness, on the contrary, doesn't have to share much with conscious experience that we have now, because its main substrate is not neural activity but individuals and interactions between them. As a result, the content of emergent consciousness won't necessarily include all the ideas and experiences that individuals have, since not all these ideas end up being expressed in interactions. So the transformation will be more qualitative than quantitative.

Ben
Hmmm... And I suppose this gives a different perspective on the question of "what gets transferred around" in a network of inter-sharing minds.

175
    http://www.boston.com/bostonglobe/ideas/articles/2010/12/19/group_iq/

It may be that in a "mindplex" formed of multiple networked minds forming a distributed shared consciousness, the localized aspects of thoughts are transferred around, then leading to the crystallization of common brain-level globalized structures in multiple brains in the network (as well as mindplex-level globalized structures spanning multiple brains)... That's how it would probably work if one networked together multiple OpenCogs into a shared-consciousness system.

Alexandra

Well I see OpenCog, when fully developed, would be great for experimenting with consciousness.

Ben

I hope so! At the moment it's easier for me to think about distributed consciousness in an OpenCog context than a human context, because we have a better understanding of how OpenCogs work than brains at this point.

Alexandra

In a way the "distributed consciousness" term better captures the essence than just "shared" consciousness. You can share in a lot of ways using BMI, even by transferring thoughts from one head to another, but what's really unique about the possibilities enabled by BMI is the potential to make consciousness really distributed.

Ben

But, thinking more about emergent consciousness, Do you think that would feel like a loss of free will? Or, given that "free will[176]" is largely an illusory construct anyway (cf the experiments of Gazzaniga, Libet and so forth, do you think our brains would construct stories allowing themselves to feel like they're "freely willing" even though the emergent consciousness is having a huge impact on their choices? Or would the illusion of free will

---

[176] http://www.goertzel.org/dynapsyc/2004/FreeWill.htm

mainly pass up to the higher-level consciousness, which would feel like IT was freely deciding things?

Alexandra

Actually, ascending onto higher (emergent) levels of consciousness, if done correctly, should feel like acquiring freedom, and not illusory.

Ben

Ah, well, that's a relief!

I suppose that's why you called it the "Consciousness Singularity" (in your presentation at the H+ Summit @ Harvard).

Alexandra

If you think about it, emergent consciousness may already exist to an extent, in places where people engage in highly interactive collective information processing (e.g. Facebook). Remember that experiments provide evidence for existence of collective, or emergent (produced by interactions between members) intelligence within groups. And this intelligence does not require wiring brains together. Language is enough to establish the necessary brain-to-brain interactions. The same can hold for consciousness. Any group of people engaged in collective activities can produce this next level of consciousness as a result of their interactions.

Ben

Right – these are examples of emergent consciousness, but they're fairly limited. And when the phenomenon gets massively more powerful due to being leveraged by brain-computer interfaces and neural implants rather than just web pages and mobile phones and so forth then you get the Emergent Consciousness Singularity!

Alexandra

Yes – and if you think about it, the Consciousness Singularity, is indeed the ultimate liberation of the immortal human soul.

In our present state, as relatively individualized minds, our freedom is greatly constrained by the society we live in. Engaging in social interactions always applies some constraints on our freedom, often very significant ones. Belonging to society makes us obey various laws and traditions – for example, not too long ago every woman had to be a housewife, and alternate careers were prohibited. Most women had nothing to do but submit to the constraint. However, the society itself as a whole was still "thinking" – in form of political debates and revolutions – on the topic of women rights, and finally got to the conclusion that set the women free. You see, individual women were not free to decide their career for themselves, due to social pressure; but society as a whole is free to think and make decisions on what it would want to look like. Hence, emergent consciousness has more freedom because it has nobody else around to set rules and constraints.

Ben
Until the emergent uber-mind makes contact with the other, alien emergent consciousnesses, I suppose... Then you'll have society all over again!

But of course, then, the various emergent consciousnesses could merge into an even bigger emergent consciousness!

Alexandra
And it will be great to live to see that sort of thing happen

Ben
Well, you've also mentioned a connection between emergent consciousness and immortality, right?

Alexandra
Exactly! If new technologies will enable us to somehow rise to this higher level of (emergent) consciousness, then we'll ultimately become not only much more free, but essentially immortal, as a bonus. We will live as long as society exists. Just like replacing individual neurons in our brain doesn't affect our

existence as a whole, similarly we won't be affected by the mortality of individuals anymore.

One may think that we'll lose ourselves in that process, but losing yourself is a necessary feature of any transformation – including such everyday activities as learning, training, acquiring new experience, and so on. And, in case of emergent consciousness, we'll still have our old self as a part of our new self, for some time.

Ben

Hmmm – yeah, about "losing ourselves". You say "we" will become almost immortal in this process. It seems clear that *something* that had its origin in human beings will become almost immortal, in this eventuality. But let's say you and me and a few dozen or billion of our best friends merge into an emergent meta-mind with its own higher level consciousness. The immortality of this meta-mind won't really be much like immortality of you or me personally, will it? Because what will be immortal will NOT be our individual selves, not the "Alexandra" or "Ben" constructs that control our brains at the moment. Although in a sense our personal streams of qualia might still exist, as threads woven into a much greater stream of qualia associated with the meta-mind.... So I wonder about your use of the pronoun "we" in the above. Do you really feel like this would be "your" immortality?

Alexandra

Sure it will be mine, but "me" will be transformed at that point -- I won't be the same "me" as before. And you and your friends, too; we all will transform from being separate entities into one big single entity. So it'll be everyone's immortality – just in case you were worried about other people being left out!

Ben

Well, I'm certainly convinced these would be interesting eventualities! I'm not quite convinced that it would be a transformed "me" living on in this sort of emergent consciousness, as opposed to some sort of descendant of "me." But actually we don't even have a good language to discuss these points. So maybe this is a good time to bring the conversation back toward the practical.

I wonder – in terms of practical R&D work going on today, what strikes you as most promising in terms of eventually providing a practical basis for achieving your long term goals?

Alexandra

I can tell you which research areas I consider the most promising in terms of providing a sufficient theoretical basis for moving on to practical developments. These research areas are complex systems science, network science and connectomics (e.g. the Human Connectome Project[177]).

Research in these areas will help us understand how to engineer a neural implant that will integrate itself successfully within a complex network of neurons; how the neural implant can fit well within the complex, self-organizing system that is the brain; and how separate, autonomous neural networks such as brains of different people can be wired together so that they will work as a system with single, unified consciousness.

Ben

Definitely an understanding of complex networks will help a lot, and may even be critical. But what about the hardware and wetware aspect? How do you think the brain-computer interfacing will actually work, when your ideas are finally realized? A chip literally implanted into the brain, with the capability for neural connections to re-wire themselves to adapt to the chip's presence and link into the chip's connectors?

---

[177] http://www.humanconnectomeproject.org/

Alexandra

Of course, it's not easy to talk in concrete details about the future given that some new technology that changes everything can emerge anytime. But I'm happy to share a few guesses.

A common guess is that some technology will be invented that will enable machines to communicate wirelessly and simultaneously with a vast number of single neurons (and their synapses) in the brain. Such technology will be a kind of ultimate solution for the BMI problem and will be useful for consciousness-expanding endeavors too. Then, we won't even need to implant anything – just wearing a special cap will be enough.

But I doubt something like that will be invented in near future.

My guess is that novel brain-machine interface designs that are applicable to my consciousness-expanding project will emerge from synthetic biology. Neural implants won't be silicon-based microprocessors like we have now; rather, they will be completely novel artificially-designed biological machines.

Neural implants made of biological substrate will integrate easily into surrounding network of biological neurons. And by applying synthetic biology tools to this substrate, we'll be able to engineer those functions that we need our implant to implement. We'll be able to tinker with the substrate so it will perform faster and better than analogous biological processors (like brains) designed by nature. And we'll be able to make this substrate compatible with classical silicon-based hardware.

Such a neural implant may be used as an autonomous brain-enhancer -- or may serve as an intermediate layer between brain and computers, being compatible with each of them!

Ben

Hmmm... I'm wondering now about the interface between the implant device and the brain. Do you think you could just insert an implant, and let the brain figure out how to interact with it? Or will we need to understand the precise means via which intelligence emerges from neural wiring, in order to make a richly functional BCI device?

Alexandra

I think it will be a compromise solution. New solutions for brain-machine interfacing will take advantage of innate brain capabilities for learning, plasticity, and self-organization. But for this, new BMI devices have to be designed in a much more brain-friendly way than now, and this will require better understanding of the brain itself. Probably, we won't need a complete description of how neural wiring result in intelligence and consciousness – but some understanding of this process will be a necessary prerequisite for good BMI designs.

Ben

Yes, I see. That sounds reasonably believable to me, in fact...

I'll be pretty happy when this sort of technology leaves the realm of frontier science and enters the realm of everyday commodity! One day perhaps instead of just an iPhone we'll have an iBrain, as I speculated once before[178]!

So maybe, when the time comes, I'll just be able to plug my iBrain into the handy cranial interface tube in the back of my head, let it burrow its way into my brain-matter appropriately, and then jack into the shared consciousness matrix, and feel my petty individual self give way to the higher emergent uber-mind.

Of course, there's a lot of R&D to be done before we get there. But it's thrilling to live at a time when young scientists are seriously thinking about how to do these things, as a motivation

---

[178] http://hplusmagazine.com/2010/12/02/iphone-ibrain-ben-goertzel-chats-humanity-director-and-mobile-app-designer-amy-li-about

for their concrete research programs. O Brave New World, and so forth...

Anyway, thanks for your time and your thoughts – I'm really looking forward to seeing how your research in these directions progresses during the next years and decades. Though of course, not as much as I'm looking forward to the Consciousness Singularity!

Giulio Prisco:
Technological Transcendence

Interview first published February 2011

A character in Ken MacLeod's 1998 novel *The Cassini Division* refers to the Singularity as "the Rapture for nerds" (though it should be duly noted that in that novel the Singularity occurs anyway!). This represents a moderately recurrent meme in certain circles – to denigrate transhumanism by comparing it to extreme religious notions. But not all transhumanists consider such comparisons wholly off-base. While transhumanism differs from traditional religions in being based around reason more centrally than faith, it does have some commonality in terms of presenting a broad vision of the universe, with implications on the intellectual level but also for everyday life. And it does present at least some promise of achieving via science some of the more radical promises that religion has traditionally offered – immortality, dramatic states of bliss, maybe even resurrection.

A host of transhumanist thinkers have explored the connections between transhumanism and spirituality, seeking to do so in a manner that pays proper respect to both. One of the most prominent among these has been Giulio Prisco, an Italian physicist and computer scientist, who is the author of the much-read transhumanist/spiritual essay "Engineering Transcendence", and the leader of the online transhumanist/spiritual discussion group *The Turing Church*. It was largely due to Giulio's gentle prodding that last year I wrote my little book *A Cosmist Manifesto*, summarizing my own views on futurist philosophy, including musings at the transhumanism/spirituality intersection. Giulio wrote a brief review/summary[179] of that book for H+ Magazine. So last month

---

[179] http://www.hplusmagazine.com/editors-blog/cosmist-manifesto-advocacy

when I decided to interview Giulio for the magazine, it was just the next step in our ongoing conversation!

Giulio has a rich history in the transhumanist community, including a period as a Board member of the World Transhumanist Association (the previous incarnation of Humanity+ [http://humanityplus.org], the organization behind this magazine), and a brief stint as Executive Director. He's also a Board member of the Institute for Ethics and Emerging Technologies[180] and the Associazione Italiana Transumanisti. And in addition to his passion for transhumanism, he has pursued a very successful career in science and engineering, including a position as a senior manager in the European Space Administration, and his current work as a software and business consultant.

Ben

OK, let's jump right into it! Could you sum up in a few sentences why you think the intersection of transhumanism and spirituality is so important?

Giulio

A few years ago I summarized my first thoughts about the intersection of transhumanism and spirituality in an article on "Engineering Transcendence", which has been published online in many different places. A few months ago I have drafted a list of Ten Cosmist Convictions, which you have substantially edited, improved and included your great Cosmist Manifesto. These are my convictions about the intersection of transhumanism and spirituality.

In summary: Our universe is a very big place with lots of undiscovered and unimagined "things in heaven and earth"

---

[180] http://www.google.com/url?q=http%3A%2F%2Fen.wikipedia.org%2Fwiki%2FInstitute_for_Ethics_and_Emerging_Technologies&sa=D&sntz=1&usg=AFQjCNEVzoUpj1Yqlhy7a3SEa0ZsYLOF_Q

which science will uncover someday, and perhaps in this mysterious and beautiful complexity there is room for spirituality and even for the old promises of religions, such as immortality and resurrection.

Or perhaps not: The Cosmist Convictions are not predictions but objectives that we may achieve sooner, later, or never. I don't know if superintelligence, uploading, immortality and the resurrection of the dead from the past will ever be technically possible. But I think these concepts are basically compatible with the laws of fundamental physics, and contemplating these possibilities can make us happy which is a good and useful outcome in itself.

Perhaps our wildest transhumanist dreams will never be achieved, but we should do our best to achieve them. Even if death is unavoidable and irreversible, we can still be happy by seeing ourselves as a small part of a big cosmic computation which, in some sense, exists beyond time, has existed, will exist, has existed in the future, will exist in the past... Death is not really a big deal, but we must do our best to cure and conquer it and become immortal... Or die trying.

I am a theoretical physicist by training, and my worldview is strictly materialist with no room for the supernatural. Actually, I am quite a pragmatist and a positivist: I don't know, or care, what "Reality" is, let alone "Truth". I only know that, under certain conditions, some models give better predictions than others and help building more useful machines. Therefore I value scientific and philosophical views not on the basis of their "objective truth", whatever that is, but on the basis of their utility. I don't have problems accepting my own interpretation of spiritual transhumanism, because I don't have conflicting a-priori metaphysical convictions.

Following William James, one of my favorite thinkers, I think a spiritual worldview based on the contemplation of transcendence can help us living better, happier and more productive lives, which is what really matters. But transcendence must be something that we build for ourselves, with science and technology.

Ben

What's the basic idea behind "The Turing Church" as you conceive it? What is it now? What would you like to see it grow into, in your wildest dreams?

Giulio

The Turing Church is a mailing list about the intersection of transhumanism and spirituality, which I have started in 2010. The name is a homage to Alan Turing, a great man killed by a bigot society, and a reference to the Turing-Church conjecture: under certain definitions and assumptions, a computer can fully emulate another computer, hence a mind, which is software running on a specific computing substrate, can in principle be transferred to another computing substrate (mind uploading). So the Turing Church can be seen as a spiritual community (Church) based on mind uploading, which is a central theme in the Ten Cosmist Convictions.

The mailing list goes through periods of activity and inactivity. At this moment I am not pushing very hard to transform the Turing Church into a "community", let alone a "Church", because there are already a few excellent communities for spiritually inclined transhumanists. So, probably the Turing Church will remain just a discussion group in the next future.

Ben

What about the Turing Church Workshop you organized last year, where I gave a brief talk on Cosmism? I'm sorry I had another obligation that day and didn't get to participate in all the discussions – but I watched rest of it on video!

Giulio

Yes, in November I organized "The Turing Church Online Workshop 1" – and all the talks and discussion at the Workshop have been recorded in full video and are available online. The Workshop took place in the Teleplace virtual world, and was populated with interested observers and representatives of various spiritual transhumanist groups, including Martine Rothblatt of Terasem, Mike Perry of the Society for Universal Immortalism (SfUI), and Lincoln Cannon of the Mormon Transhumanist Association (MTA)... And a guy who wrote a Cosmist Manifesto last year, who gave a great talk without really answering the question I asked him to address in his talk: how to bootstrap a "Confederation of Cosmists".

Ben

Hah. Let's get back to the Confederation of Cosmists a little later, and talk a bit about the Workshop now.

Giulio

Sure! I also gave a talk on "The Cosmic Visions of the Turing Church", similar to the talk I had given at the Transhumanism and Spirituality Conference 2010 hosted by the MTA and partly based on my article "In Whom we live, move, and have our being". In my presentation I developed the idea of scientific resurrection: our descendants and mind children will develop "*magic science and technology*" in the sense of Clarke's third law, and may be able to do grand spacetime engineering and even resurrect the dead by "*copying them to the future*"... Of course this a hope and not a certainty, but I am persuaded that this concept is scientifically founded and could become the "missing link" between transhumanists and religious and spiritual communities. I am perfectly aware that I can be accused of building a castle of unfalsifiable pseudoscience to protect my wishful thinking from reality, and I can live with that.

The objectives of the Workshop included discovering parallels and similarities between different organizations and to agree on common interests, agendas, strategies, and outreach plans. I think these objectives have been achieved, and some participants who were initially members of only one group have also joined the others, which is a good sign. I am a member of the three organizations, and I see them as compatible and complementary, each with its own specific focus.

Firstly, the Mormon Transhumanist Association is a very interesting experiment in promoting transhumanist spirituality within a major religion. Being a Mormon is not required for membership in the MTA, but of course most members are Mormons. Lincoln Cannon and others say that the Mormon doctrine make it very compatible with transhumanism, and I find their arguments persuasive.

Ben

Yes, I actually talked with Lincoln extensively earlier this month, and I interviewed him for H+ magazine – the interview should appear not to long after yours.

Giulio

The Mormon transhumanists' New God Argument (a revised version of the simulation theories of Moravec and Bostrom) is especially relevant to spiritual transhumanism. They have made some friends in the Mormon community at large, but of course also some enemies, and I have my doubts that their views may be accepted by the more socially retrograde segments of the Mormon community. However, this is a great group and I am proud to be part of it. I look forward to seeing similar initiatives within other Christian denominations and religions.

Ben

Their thinking is certainly interesting, though I have to say I find much of it unconvincing myself. Still, I came away from my conversations with Lincoln incredibly impressed with his creativity, and with the ability of traditional human belief systems like Mormonism to stretch and bend to account for transhumanist ideas. He has genuinely managed to interpret Mormonism in a way that's consistent with radical transhumaism, and that's a fantastic intellectual achievement. It makes me feel that, even after technological advancement brings really radical – even Singularity-level – changes, many people will still find no problem incorporating these in (upgraded versions of) their traditional belief systems. Though, I do have a suspicion that after a transitional period these traditional belief systems will fall away – even Mormonism!

Giulio

Another group represented at the Workshop was the Society for Universal Immortalism. While this group is very small and it has not been very active recently, I am very fond of it because it gives a central role to the idea of technological resurrection which as I say above is, I believe, a necessary component of transhumanist spirituality. Universal Immortalism is built on Mike Perry's book "Forever for All", a great book which is now also featured on one of the main Terasem sites, the Truths of Terasem Podcasts site maintained by Alcor founders Fred and Linda Chamberlain, who are now very active in Terasem and participated in the Workshop.

Ben

Terasem is interestingly diverse – I gave a talk at one of their workshops in Vermont a couple years ago, which was nominally about nanotechnology and its implications, but actually touched on all sorts of transhumanist technologies and issues.

Giulio

Terasem is a galaxy whose center is everywhere and nowhere or, in the words of its founder Martine Rothblatt in a recent email, in *"all the anarchic Terasem stuff on the web"*. At the Workshop, Martine described how the Terasem idea was born in her mind in 2002 with a powerful serendipitous epiphany of *"balance of diversity and unity, in a quest for joyful immortality"*.

Ben

So are there any plans for another Turing Church Workshop?

Giulio

I will think about the future of the Turing Church and probably organize a second Workshop, but at this moment I am more inclined to participate in already existing groups.

Ben

You have a Turing Church discussion group, and then there are the Workshops. Do you see there as being actual Turing Church "services" at some point in the future, similar to what happens now at Christian churches, or Buddhist temples, etc.? Do you think these might occur in virtual worlds as well as in the physical world? What elements do you foresee might go into such a service – speculating freely!

Giulio

As I said, I am not planning to develop the Turing Church into more than a discussion group. I may change my mind, but at this moment I prefer being an active member of existing spiritual groups. I am not really interested in becoming a leader, or a guru.

There have been discussions of "services" in the Universal Immortalism mailing lists, but I don't think any firm plan has been made. Of course Mormon Transhumanists participate in Mormon services, but I am not aware of any services for the transhumanist subgroup.

Terasem has services which, in line with the distributed and decentralized nature of the organization, can be performed anywhere with any number of participants (even one). Terasem services have a New Age look & feel, and I think they can help building bridges to the large New Age community. Some Terasem services have been performed in Second Life, and I look forward to participating. The Mormon Transhumanist Association is also hosting a monthly discussion group in Second Life.

Giulio Prisco: Technological Transcendence

A few years ago we founded an "Order of Cosmic Engineers" with a mailing list and periodic events in Second Life. Some spinoffs of the Order of Cosmic Engineers are active, and so is the original mailing list. I still participate sporadically, but I am mostly focusing on other things now.

I wish to recommend a great book: "Apocalyptic AI: Visions of Heaven in Robotics, Artificial Intelligence, and Virtual Reality" by Robert Geraci, who also participated in the Turing Church Online Workshop. Here is my review of Robert's book. Chapter 3 is entirely dedicated to his long field recog mission behind transhumanist lines in Second Life. He attended the 2007 Seminar on Transhumanism and Religion in SL and many events related to transhumanism and religion in 2008 and 2009. In the process, he became a friendly observer in our community, and a good friend of many transhumanist users of Second Life.

From my review:

> "According to Geraci, Apocalyptic AI is a religion: it is a religion based on science, without deities and supernatural phenomena, but with the apocalyptic promises of religions. And he thinks that, while the Apocalyptic AI religion has a powerful but often hidden presence in our culture, the Transhumanist community embraces it openly and explicitly. Transhumanism is first defined as "a new religious movement", and throughout the book Geraci continues to see it as a modern religion. This may shock and upset many transhumanists readers who proudly see themselves as champions of science and rationality against religious superstition. Not this reader, though. I remember my first impressions after joining the Extropy mailing list in the late 90s. I thought, this is a powerful new religion for the new millennium."

As you know, I have been personally and professionally involved in Second Life and other virtual worlds for years. Now I am focusing on other technologies to make a living, but I still have a few avatars in Second Life and look forward to participating in talks, seminars, conferences and why not religious services. I am an advocate of <u>telepresence solutions</u> for business, education and entertainment, and why not spirituality. It is clear that virtual reality technology can only improve, and I think virtual worlds can be very a very powerful enabling technology for new spiritual movements.

There is a caveat though. For most people virtual worlds are just nice things that one can do without, but a few users find them very addictive. Some of these users may become exclusively focused on their avatars or "digital persons", often because in virtual worlds they can easily correct shortcomings of their real life which, unfortunately, cannot yet be easily corrected in the real world. I have come to consider some of these digital persons as close friends.

Now, I am a radical libertarian as far as personal lifestyle choices are concerned. I am also persuaded that in the future many people will upload to cyberspace, and I see this as something good. So, I am the last person who would object to the aspirations of my digital friends. But our primitive (by future standards) virtual reality technology is not yet able to permit living an alternative life in the metaverse. Someday it will be, and we will flock to cyberspace leaving our meat bodies behind. But that day is not now, and I am afraid addiction to virtual worlds can lead to stop *even trying* to function in the real world, which can be psychologically damaging. I guess what I am trying to say is, use virtual worlds as powerful telepresence and communication environments, and use them as much as you like for fun and role-playing, but please don't forget that they are not real worlds yet. Someday, but not now. I am very happy to see that a good friend (you know who you are) is now stepping out of the Second Life greenhouse and becoming more and more active in the real world.

Giulio Prisco: Technological Transcendence

Ben

Your current choice to pursue the Turing Church meme via discussion forums and workshops, rather than more traditional church-like stuff, brings up another point I was wondering about. I'm curious what's the difference, in your view, between Transhumanism or Cosmism as *philosophies* (as you'll recall I've referred to Cosmism as a "practical philosophy") and your idea of a transhumanist or cosmist *religion*. That is, what elements do you see a religion as having that a philosophy lacks, and why do you feel these elements are important?

Giulio

Religions are huggable: They help you get through the night and give you some warmth in a cold universe. Practical philosophy is also huggable. I like Cosmism as a practical philosophy because I think it can help me living a better and happier life, and I liked your book much more than most abstract philosophy books.

I like to play with abstract ideas, and so I have spent a lot of time reading philosophy. But besides a certain intellectual fun I don't find most of philosophy very relevant. Abstract philosophy tends to be concerned with things like "Objective Reality", "Objective Truth", and "Objective Good", but I am not persuaded that such things exist, and I am not persuaded that we really need them. The important questions I ask myself and others are more like "Can this improve my life?", "Can this make my family, my friends and the other people on the planet happier?" I think if something can make some people happier without making others less happy, then it is definitely good and no amount of abstract sophistry can make it less good.

Cosmism in the sense of practical philosophy as described in your book, the spiritual visions of the Russian Cosmists including Fyodorov's radical ideas on immortality and resurrection of the dead (described in more modern terms by Mike Perry in "Forever for All"), Hans Moravec's and Hugo DeGaris' visions of a Cosmist diaspora with ever increasing intelligence spreading outwards from the Earth into the cosmic night, and the wild speculations in the Ten Cosmist Convictions, can give us joy and help us becoming better and happier persons.

I am persuaded that a wonderful cosmic adventure awaits our descendants and mind children, and I wish to be part of it. Perhaps radical life extension and mind uploading will be developed soon enough for me, or probably not. Perhaps cryonics will help, and perhaps not. Perhaps future generations will be able and willing to retrieve me from the past and upload me to the future as part of an artilect who, sometimes, may remember having been me... Or perhaps not. But I am persuaded that these glimpses into veiled future possibilities are basically compatible with our best understanding of how the universe really works. And they give me beautiful visions of future worlds, and the hope, drive and energy that I need to try living a good and productive life, in this world. And this, I believe, is what really matters.

Ben

What has been the reaction in the transhumanist community to your ideas about joining together transhumanism and spirituality? What has been the reaction in the traditional spiritual, religious community? Has there been enough positive reaction on both sides to give you the feeling that Turing Church type initiatives have real potential to grow into something important in the world?

Giulio

The reaction in the transhumanist community has not been very enthusiastic, but that was expected because many transhumanists force themselves into a hard ultra-rationalist attitude and can hardly swallow soft "possibilian" approaches. I think in many cases they had a very hard time overcoming their traditional religious upbringing, and they are constantly afraid of falling back. I was not raised in a very religious family and, apart from a couple of very mystic weeks when I was 14, I have never taken religion too seriously, so I am not afraid of falling back anywhere.

There is a small but growing minority of transhumanists who like my ideas. I am also proud to see that I am frequently attacked by anti-transhumanists for my ideas about joining together transhumanism and spirituality. Sometimes anti-transhumanists see the nature of our ideas more clearly than transhumanists themselves, and they consider it as a threat to conventional religions in their own space. Since I started reading transhumanist thinkers in the early 90s, I never had any doubts that our worldview is a religion: We don't believe in the supernatural or in a dogmatic, revealed faith, but we are determined to make the old promises of conventional religions come true by means of science and technology. We see the universe with a sense of wonder and a sense of meaning, and our beautiful visions make us hopeful and happy. In this sense, transhumanism is a religion, and it has always been one.

I would not know about the reaction in traditional spiritual, religious communities, because I am not a member of any. When I describe my ideas to religious persons, they are usually shocked at the beginning but often they begin to listen with interest at some point. My friends of the Mormon Transhumanist Association are, of course, part of the Mormon community, and I believe their experience is similar.

I believe we should try to develop closer relations with New Age groups. New Agers have a bad reputation in the transhumanist community where they are seen as hopelessly deluded fools who waste their time jumping from a crystal therapy to a pyramid energy healing group, and are easy prey of new gurus and new scams, but I see them as fellow travelers and seekers, who could be more interested in our spiritual ideas than ultra-rationalist transhumanists or dogmatic religious believers. One of the reasons why I am very fond of Terasem is that I think it could be very appealing to New Agers.

Ben

It's not surprising if the reaction to your ideas has been all over the map, since the transhumanist community itself is extremely diverse and generally all over the map!

Giulio

Indeed. In this discussion I have used frequently the expression "transhumanist community" because it was kind of required by the context, but in reality there is no such a thing as a transhumanist community.

There are self-identified transhumanists all over the political spectrum from the far right to the far left, there are libertarian and authoritarian transhumanists, there are all sorts of positions on important social and political issues, and there are all kinds of attitudes toward the spiritual dimensions that we have discussed, from religious believers to intolerant "New Atheists".

The common denominator, that using advanced technologies to substantially modify humans is feasible and desirable, does not hold transhumanists together any stronger than, say, members of a science fiction salon. Then, I think it is time to stop pretending that there is a "transhumanist community" and openly acknowledge that there are many separate groups, with similarities but also important differences. Once we acknowledge this, we may be able to work together on specific projects, openly or behind the scenes.

Based on this perspective, I've personally decided that I will dedicate less time to promoting a generic concept of transhumanism, and more time to more specific projects and groups. In particular, I want to be more actively involved in technology development, I want to help developing Cosmist ideas, and I want to participate more actively in the spiritual transhumanist groups mentioned here.

Ben

One concept that seems a help a bit in disambiguating the multiple varieties of transhumanism is the distinction between "weak transhumanism" and "strong transhumanism" – where the latter is a more thoroughgoingly transhumanist philosophy, that verges more on Cosmism and is more markedly distinct from traditional humanist philosophy. We've talked about this before. I wonder if you could elaborate on this weak vs. strong transhumanism distinction and how you see it tying into your thinking on the transhumanism/spirituality intersection.

Giulio

If weak transhumanism is about mild anti-aging therapies paid by the public health system, and strong transhumanism is about mind uploading to cyberspace and galactic colonization, then of course I am both a weak and a strong transhumanist.

But the distinctive feature of weak transhumanists is their bigot condemnation of strong transhumanists, and their thought-policing attitude. According to them we should repent of our juvenile sins, abandon our transhumanist dreams, and think only politically correct thoughts. My short answer to that is: *"BS"*. In more words: *"Cut the crap, think with your own head, and let me think with mine."* One thing that really drives me mad is denying others the liberty to think with their own head.

As far as I am aware, no strong transhumanist has ever condemned weak transhumanists for preferring to focus on proximate here-and-now concerns. But weak transhumanists do condemn radical and visionary transhumanists, and so we all have to choose a camp. I have chosen mine: I am a visionary, imaginative, irreverent, far-future oriented, politically incorrect transhumanist.

At the same time I have to mention that, besides their annoying bigot attitude, I quite frequently agree with weak transhumanists or even anti-transhumanists. I share many of their views on contemporary social and political issues, and some of my "enemies" would be surprised to learn that I agree with them on almost everything. I don't think strong AGI and mind uploading will be developed soon, I think public research funding should give priority to more proximate projects, and I don't believe in a clean singularity because the real world is messier than our nice models. In general, I think Ray Kurzweil's predictions are far over-optimistic.

But Kurzweil gives us good dreams to dream, and often good dreams can be turned into reality. This is Ray's role, and one Ray is worth thousands of his detractors. This is strong transhumanism: We dream wild dreams, and we do our best to make them come true.

In your Cosmist Manifesto, you point out that we can only achieve great goals by putting ourselves in a positive state of mind. Before a competition, a great athlete is in a state of flow and sure to win. Of course, in their rational minds great athletes know that victory is not certain and many things can go wrong, but positive thinking helps them giving their very best. It is in this sense that we, radical transhumanists, prefer an optimist outlook. It is in this sense that we say that we will roam the universe beyond the most remote galaxies, overcome all human limitations, and achieve transcendence.

Strong transhumanism is inherently spiritual, and Cosmist. Science and technology are not enemies of our desire for transcendence, but on the contrary they are the very means which will permit achieving transcendence.

Ben

Hear, hear!

Finally, as a reward for answering all my interview questions so wonderfully, I'll make a brief effort to answer your questions about how to start a Confederation of Cosmists – i.e. some sort of active organization centered around Cosmism aka "strong transhumanism" as a practical philosophy. Not a religion but something vaguely analogous to a religion, founded on rational radical transhumanism rather than traditional faiths.

The reason I didn't address that in my talk at the Workshop was basically that I didn't have any really exciting or original answers. And I guess I still don't.

Giulio

But now you will share your boring and unoriginal answers?

Ben

Precisely My thinking is that there are two good ways for such a Confederation to get founded in the near future.

The first is simply the emergence of a guru. Someone who wants to lead such an organization, and go around giving speeches and posting videos and writings about it, and being the charismatic leader at the heart of it. That could certainly work – the affinity of humans, even transhumanist humans, for charismatic leaders seems undiminished as technology advances.

The other, perhaps more interesting route, would be to somehow leverage social network effects. Make a Cosmist Confederation that could grow virally and self-organizingly in the same sort of way that Facebook, LinkedIn and Twitter have grown. Note that those social networks don't have any charismatic leaders at the center. Now we don't have any example of religion-type organizations that have self-organized in this manner. But of course the Web and other modern communication technologies are new, and it's to be expected that all sorts of new phenomena are going to emerge in the coming years and decades.

This brings me back to an idea I had in the late 1990s, before blogging as such existed – I wanted to make a massive, collectively-authored, web-based Knowledge Base of Human Experience. The idea was that each person could log on and type in some text about an experience they'd had, maybe upload some relevant images, and then enter some meta-data describing the emotional or spiritual or social aspects of the experience, to enable effective automated search and indexing and organization. I never built the website unfortunately – if I had maybe I would have gradually morphed it into what we now call a public blog site, and gotten rich... But I wonder if some idea vaguely like that could work today, and could serve as the seed of a self-organizing network-ish Cosmist Confederation?

This ties into something I've often complained about regarding the transhumanist community (I know you say that community doesn't really exist, but as you say it's a useful shorthand!) – the lack of attention to inner experience. Too much about the science and technology, not enough about the individual and social states of mind that result from the technology, and that ultimately enable and drive the technology. Perhaps some sort of Web social network specifically oriented toward sharing states of mind and experiences related to technology, mind and spirit could grow into something interesting. What if you could upload an experience you've had, and get automagically connected with others who've had similar experiences? Well you can envision the possibilities...

I don't have any details brilliantly worked out or anything. But my general point is, somewhere in the intersection of accelerating change, subjective individual and social experience, and Web 2.0 tech, there probably lies a way to seed the self-organization of a Confederation of Cosmists, without the need for any guru at the center. If I weren't so busy trying to build superhuman thinking machines and save the world, I might be inclined to figure out the right way to design that social network... But as it is I'll be happy to leave it to somebody else and sign on as an enthusiastic user. Or if some industrious reader of this interview wants to collaborate on such a thing, who knows, maybe something like this could happen.

Zhou Changle: Zen and the Art of Intelligent Robotics

Interview first published February 2011

In the last few decades, science has unveiled a tremendous amount about how the mind works – exposing various aspects of the hidden logic of cognition, perception and action, and connecting the mind to the brain. But modern science is not the first discipline to probe the nature of the mind. Multiple traditions within diverse cultures throughout history have probed the mind in their own ways. For example, my teenage self spent many perplexed hours poring over books on Buddhist psychology, reviewing the theories of the medieval Indian logicians Dharmakirti and Dignaga, whose detailed analysis of the different possible states of consciousness seemed in many ways more revealing than modern consciousness science.

Among the numerous approaches to the mind, Zen Buddhism ranks as one of the most intriguing. Far less theoretical than modern science or medieval Indian logic, Zen emphasizes experiential wisdom and the attainment of special states of consciousness – Zen "enlightenment." While a great deal has been written about Zen, this is a literature fraught with irony, since one of the central notions of Zen is that true understanding of the mind and world can't be transmitted in language.

According to the traditional story, Zen began one day when Buddha, addressing an audience gathered to hear his teachings, chose to remain silent instead. Instead of discoursing as expected, he stood there quietly holding a flower in his hand. Most of the audience was confused, but one disciple, Kashyapa, experienced a flash of sudden understanding – and gave the Buddha a subtle smile. Buddha gave Kashyapa the flower. Then Buddha finally spoke to the crowd, "All that can be given in words I have given to you; but with this flower, I give Kashyapa the key to all of the teachings." This is the origin of Zen – a "wordless transmission" of mind to mind. And – to continue the traditional tale – 1300 years or so later, in 500AD, the Indian monk Bodhidharma brought the wordless teaching to China, founding Chan, the Chinese version of Zen which later spawned the somewhat different Japanese version of Zen, that is perhaps the best known species of Zen in the West.

Zen is famous for brief riddles called *koans*, such as "What is the sound of one hand clapping?" Meditating on such riddles has the potential to bring enlightenment, when the mind suddenly realizes the futility of approaching the koan (or the world or the mind) with conceptual thought, and fully embraces the paradox at the heart of the koan (and the world and the mind).

Chinese Zen spawned what has been my favorite statement about Zen since I first encountered it when I was 18 years old, in the wonderful little book The Zen Teachings of Huang Po[181]:

> *The dharma of the dharma is that there are no dharmas, yet that this dharma of no-dharma is in itself a dharma; and now that the no-dharma dharma has been transmitted, how can the dharma of the dharma be a dharma?*

The word "dharma" means, roughly, the essential teachings of the Buddha (which lead to enlightenment) and the essential constituent factors of the world. Huang Po gives us Zen straight up, not prettied-up for Western consumption like some of the Western Zen literature (e.g. Robert Pirsig's famous and excellent book Zen in the Art of Motorcycle Maintenance[182] which inspired the title of this interview).

And Zen is alive and well in China today, as I found out for myself recently. During the period 2009-2011 I enjoyed

---

[181] http://www.amazon.com/Zen-Teaching-Huang-Po-Transmission-Mind/dp/0802150926

[182] http://www.amazon.com/Zen-Art-Motorcycle-Maintenance-Inquiry/dp/0061673730

collaborating on AI and robotics research with a lab at Xiamen University in south China; and on one of my visits to Xiamen, I was delighted to learn that Zhou Changle, the dean of Cognitive Science at Xiamen University, is a very accomplished Zen practitioner.

Changle was one of the original masterminds of the Artificial Brain project that brought my good friend Hugo DeGaris to Xiamen a few years ago. Hugo retired recently, but the project continues within Changle's BLISS (Brain-Like Intelligent SystemS) lab. They have 3 humanoid Nao robots in the BLISS lab, together with some supercomputing equipment and a host of capable grad students and professors, and are working on making the robots intelligent and (in at least some sense) self-aware.

Hugo's main interest, in the context of the Xiamen project, regarded large-scale brain simulation using neural nets on specialized hardware. On the other hand, Changle's scientific contribution to the project has more to do with machine consciousness and the logic of mental self-reflection – topics that, after I learned of his involvement in Zen, seemed to fit in perfectly. And beyond our collaborative AI/robotics project, Changle also carries out research on a host of other topics including computational brain modeling, computational modeling of analogy and metaphor and creativity, computational musicology and information processing of data regarding traditional Chinese medicine.

Changle originally pursued his study of Zen under a teacher – but after several years of that, his teacher judged he had reached a sufficient level of enlightenment that he should continue on his own! And as well as practicing Zen and various forms of science, he has written extensively (in Chinese) about the relation between the two, including over a dozen papers and a book titled *TO DEMONSTRATE ZEN— ZEN THOUGHT VIEWED FROM SCIENCE.*

The abstract of one of his recent talks gives a flavor of how he connects Zen with cognitive science:

> The various states of consciousness experienced by human beings may be divided into four categories, using the philosophical concept of intentionality: Intentional, non-intentional, meta-intentional, and de-intentional states. Analyzing Zen "enlightenment" in the light of this categorization, one concludes that Zen thinking is a de-intentional self-reflective mental capacity. This establishes a philosophical basis for the Zen method of mind training, enabling the exploration of connections between Zen, philosophy of mind, cognitive science, and other areas.

However, Changle's talks are much more colorful than this conveys, filled with humorous stories in the rich Zen tradition.

When I interviewed Changle about Zen and robotics and consciousness, we focused specifically on some areas where we don't quite agree, regarding the relationship between AI and consciousness.

Changle foresees AIs and robots may eventually become extraordinarily capable, potentially even launching some sort of technological Singularity. But he also thinks they will always be impaired compared to humans in their internal experience, due to a lack of a certain type of "consciousness" that humans possess – a "consciousness" closely related to Zen enlightenment.

On the other hand, I myself suspect that AIs and robots can eventually become just as intelligent and conscious as we humans are. And ultimately, I suspect, they can become just as enlightened as any human – or more so.

Unsurprisingly, during our conversation we failed to come to full agreement on this set of issues – but we did have an interesting dialogue!

Ben

Changle, I've interviewed a number of scientists and also some spiritually-oriented people – but you're the first person I've interviewed who's BOTH an extremely accomplished scientist AND deeply experienced as a spiritual practitioner within a particular wisdom tradition (Zen). I'm curious how you balance these two aspects of your mind and life. Do you ever find them pulling you in two different directions?

Changle

Zen, strictly speaking, is not a religion in the sense that Buddhism, Christianity and Islam are. It is a philosophy towards life. Thus it won't collide with science. In the purest sense of Zen, it doesn't even emphasize meditation. Freedom of mind and enlightened wisdom are what matter.

Ben

It's an intriguingly paradoxical philosophy towards life – and its elusiveness seems part of its essence. I remember you gave a talk about Zen and science entitled "Not Two". And later I saw the Chinese characters for "Not Two" on an archway in a big Buddhist temple complex on Hainan island.

Changle

Zen basically goes against any forms of dualism. A is true; A is false; A is true and false; A is not true and not false.

Ben

So once you've described Zen in some way, you've set up a dualism between the things that match your description, and the things that don't. And so what you're describing isn't really Zen, which is non-dual. It's simple to the intuition yet perplexing to the thinking mind. I get it. And I don't get it. And I both get it and don't get it!

Heh... But I remember once you said that only Chinese could become enlightened! So I guess I don't get it.

Changle
That was a joke, of course!

Ben
Of course – but the reason it was a funny joke is that there's a little bit of truth at the core of it.

Changle
Yes, the intuitive Chinese mentality and the holistic Chinese philosophy have advantages over the Western analytic mentality of dualism, in terms of grasping the essence of Zen. But even so, Zen is still very elusive to most Chinese.

Ben
Now let's turn to your AI research, and its relationship to your Zen intuitions.

Superficially, there would seem to be a certain disharmony between the views of the mind implicit in AI research and Zen. AI is about taking parts of the material world, and arranging them in such a way that intelligent behavior results. But Zen takes a very different view of the mind – it views the mind as more like the basic ground of existence, not as something that comes out of arrangements of material objects like semiconductors or switches or neurons. How do you reconcile these two positions?

Changle
To me, the material world reveals itself through the human mind. In other words, it never exists without a human understanding it. However, mind is not an entity, but rather a process of making sense of everything. The intelligence that we are achieving through robots is not an entity either, but rather a process of demonstrating intelligence via the robot.

Ben
But what about consciousness? Let's say we build an intelligent computer program that is as smart as a human, or smarter. Could it possibly be conscious? Would it *necessarily* be

conscious? Would its degree and/or type of consciousness depend on its internal structures and dynamics, as well as its behaviors?

Changle

Consciousness, as I understand it, has three properties: self-referential, coherent and qualia. Even if a robot becomes a zombie that acts just like it possesses consciousness, it won't really possess consciousness – because the three properties of consciousness cannot all be realized by the reductive, analytic approach, whereas AI is based on this sort of approach.

Ben

Hmmm... But I wonder why you think the three properties of consciousness can't be demonstrated in a robot? I agree that human-like consciousness has three properties (self-reference, coherence and qualia) – there are many other ways to describe the properties of consciousness, but that seems OK to me. But as a working assumption, I'd be strongly tempted to assume robot can have all of these.

I wonder which of the three properties of consciousness you think a robot is incapable of having? Qualia, coherence or self-reference? And why is it that you think a robot can't have them?

To me, a human can be thought of as a kind of "meat robot". So I don't see why a robot can't have all the properties a human can have.

Though I have to admit my own view of consciousness is rather eccentric. Although I don't identify as a Buddhist, I share with many Buddhists the notion of panpsychism – the notion that "everything is conscious" in a sense.

But of course I can see the "reflective, deliberative consciousness" that humans have is something different than the universal consciousness that's immanent in everything including rocks and particles. I wonder what's your view on the relation between these two types of consciousness – on the one hand, the "raw consciousness" that we panpsychists see everywhere; and on the other hand, the "reflective, deliberative consciousness" that humans have a lot of, mice may have a little of, and rocks seem to have basically none of.

Changle

In Zen we say that SUCHNESS is the nature of everything. For a conscious human being, its Suchness is its consciousness; but for material things, their Suchness is their properties such as shape, substance or weight but not consciousness.

Ben

Hmmm... The word "suchness" seems hard to translate from Chinese to English!

Turning to the Internet for guidance, a writer named A.H. Almaas says[183] that:

> *In simple terms, to experience ourselves as Being is to experience our existence as such, to experience our own Presence, our own "suchness" directly. It is the simplest, most obvious, most taken-for-granted perception that we exist. But this existence is usually inferred, mediated through mind, as in Descartes's "Cogito ergo sum" – "I think, therefore I am." Existence is background, not foreground, for our ordinary experience. To penetrate into this background, to question our assumptions about reality and ourselves, allows us to encounter directly the immense mystery of the arising of our consciousness and of the world.*

It is Suchness, pure Suchness. We cannot say anything about it. We cannot say it is self, we cannot say it is not self, we cannot say it is God, we cannot say it is the universe, we cannot say it is a person, we cannot say it is not a person; the moment we say anything, we are within mind. If we use any concept here, even

---

[183] http://www.ahalmaas.com/Glossary/s/suchness.htm

the concept of purity, simplicity, or whatever else, we are within the mind, and we are blocking that which cannot be named.

Changle

The Chinese word for "suchness" is "真如" or "自性", i.e. "buddhahood", also known as "the foundational 识" or "种子识" or "Alayavijnana" etc.... It is the name for the nature of everything.

Ben

So you say everything has Suchness, but only for certain systems like humans does the Suchness manifest itself as consciousness per se. I'm reminded of some Western philosophers like Stuart Hameroff, who say that everything has "proto-consciousness", but only in some specific cases like people or other advanced animals does the proto-consciousness manifest itself as consciousness.

Whereas, I would say that everything has "consciousness" but only in some specific cases like people or other advanced animals *or sufficiently sophisticated AIs*, does this basic consciousness manifest itself as "reflective, deliberative consciousness."

But it's actually quite hard to tell the extent to which we disagree on this point, versus just having very different terminologies. Human language was created to discuss practical matters, not to have precise discussions about subtle points like this... And translation between English and Chinese makes it even trickier.

Changle

I'll explain my view in a little more detail.

According to the Buddhist perspective, mind is a kind of process, not a kind of entity. The process of mind contains: five kandhas (also called aggregates) and eight senses (八识) which are also often thought of as "eight consciousness."

The eight senses are
1. Eye (vision/visual sense)
2. Nose (olfactory sense)
3. Ear (auditory sense)
4. Tongue (taste sense)
5. Body (tactile sense)
6. Mind (consciousness)
7. Manas (realization/awareness)
8. Alaya (seed, prajna, suchness)

Note: the last one is the "suchness"!

If you want to compare these Buddhist concepts with "consciousness", the sixth sense ("真如", translated as Mind above) is the most similar with the western concept of "consciousness". But according to Zen, Manas (the seventh sense, i.e. the ability of awareness or realization) is more nature than consciousness; and the only thing which can decide the Manas is Alaya (Suchness). So really, the concept of "真如" doesn't exist in the system of Western philosophy – it's not exactly the Western concept of "mind" or "consciousness."

And corresponding to these eight senses, the five aggregates identified in Buddhism are

1. Matter or Form (rupa): Perception, i.e. the physical form corresponding to the first five organs of the eight senses, which has intentionality. AI robots can have this ability.

2. Sensation or Feeling (vedana): The feeling in reception of physical things by the senses through the mind or body; emotion, also known as "qualia", which doesn't have intentionality. It's hard to give AI robots this kind of ability, but probably not impossible.

3. Thinking (senjna): Thought and cognition, which has intentionality. AI robots can have this ability.

4. Volition or Mental Formation (samskara): Conditioned responses to objects of experience, such as language, behavior, ideas and so on, possessing intentionality. Robots can have this ability.

5. 识: Awareness/Realization/Subconsciousness, which is anti-intentionality-ed / de-intentionality-ed / un-intentionality-ed. And this is beyond the scope of robots!

Regarding the project of implementing the "mental power" of humans in robots, the key troublesome point is number 5 in the list – the implementation of "awareness mechanics" (i.e. the ability of de-intentionality), which is the so-called enlightenment of Zen. For this purpose, current AI research, which is based on a reductive and analytical approach, is not helpful at all.

Ben

I think I understand what you're saying; and I find this perspective fascinating. And I know the details run a lot deeper than what you've described just now; you've kept it compact and simple since this is an interview not a long lecture...

But still – as you probably expected, I'm not convinced.

Once again... Why do you say that a robot cannot have 识 (awareness, de-intentionality)?

I mean, think about people for a moment. From one perspective, humans are *meat* – we're made of organs, cells, molecules, etc. From another perspective, humans are minds that have 识 .

Similarly, from one perspective, robots are machines -- they're made of wires, memory disks, silicon molecules... But from another perspective, perhaps robots can eventually be minds that have 识, just like humans...

After all, Ben cannot directly experience Changle's 识. Just like Ben cannot directly experience a robot's 识. But even so, Changle's 识 has its own suchness, and perhaps the robot's 识 can have its own suchness as well – if the robot is made correctly! Say if it's built using OpenCog.

Changle
The awareness (enlightenment) of Zen is beyond all concepts, but all the approaches we use for building robots and AI systems, and all the behaviors of robots and AI systems, are based on concepts

Ben
Heh... I had intended to ask you sometime during this interview whether you think an AI program or an intelligent robot could ever experience enlightenment in the same sense that a human can. Can a robot attain enlightenment? – can we have a Zen Master bot? But I think your remarks have made your view on this pretty clear.

Changle
Right. Robots can't be enlightened because they don't possess consciousness.

Ben
Your view has an impressive coherency to it. But even so, I'm going to keep on being a stubbornly unenlightened Westerner! My perspective is more like this:

- We may *initially construct* an AI robot according to our plans and ideas and knowledge – our concepts, as you say.
- However, once we release the robot into the world, and it starts interacting with the environment and self-organizing, *then* the robot becomes its own being ... it goes beyond our initial programming, just as an embryo grows beyond the original cells that make it up, and just as a baby eventually goes beyond its initial neural wiring. Through its own growth,

self-organization, development and interaction with the world, the robot's "suchness" can develop consciousness, insight, human-like qualia and knowledge over time.

At least, that is my current belief and hypothesis. But we will need to build MUCH better AI systems and robots to find out if I am correct or not!

And of course, that's what we're working on together in the BLISS lab in Xiamen…

Anyway it's good to hear your different perspective, articulated so well. Life would get boring if everybody agreed with me…

I don't think we're going to get much further on the question of consciousness just now, so let me shift attention to a slightly different question, still in the robotics vein. I wonder just how sensitively you think humanlike intelligence and consciousness depend on having a humanlike body.

For instance, in our lab in Xiamen we have 3 Nao robots, which our students are controlling with AI algorithms. The Nao robots are roughly humanoid looking, but they have many shortcomings relative to human bodies – for instance, their skin has no sensation in it; and they can't feel any sensations inside their body; they can't move their faces in accordance with their emotions; etc. Setting aside subtle issues of consciousness – do you think these physical shortcomings of the Nao robots will intrinsically prevent an AI embodied by the Nao robot from having a human-level intelligence? How realistic do you think a robot body would have to be to really support human-like intelligence?

Changle

The human intelligence and body are not independent of each other. If they are viewed separately, that is dualism. The robot's AI and the robot's body are not independent of each other either.

Ben

OK – so one last thing, then. You're familiar with Ray Kurzweil and Vernor Vinge's notion of the Singularity – a point in time at which technological change becomes so rapid that it's basically infinite from the human perspective, and superhuman AIs rather than humans are the main driving force on the planet. Kurzweil foresees it will come about in 2045. What do you think? Will a Singularity happen this century? Will a Singularity happen at all? If a Singularity does happen, what will it mean in Zen terms? Could a Singularity possibly mean Enlightenment for everyone, via for example special computer chips that we plug into our brains and that stimulate the brain's spiritual centers in a way that spiritually "wakes a person up"?

Changle

Such technology breakthroughs are possible. But putting computer chips in the brain might not produce novel conscious experiences, if consciousness is discovered to be based on quantum entanglement.

Ben

Hmmm... Once again you remind me of Stuart Hameroff a bit – what he calls proto-consciousness is closely related to Zen "suchness", and he's quite convinced that human consciousness depends on macroscopic quantum phenomena in the brain.

However, again, I'm a bit unconvinced. I think there are some subtle points about quantum entanglement that no one really understands yet. For instance, there's some mathematics suggesting that quantum entanglement is really about the relationship between a system and the observer, not just about the system in itself. So, in some cases a system typically modeled using classical physics, may be best modeled (with respect to a given observer) using quantum mathematics and quantum entanglement. I wrote a blog post[184] about this a while back; maybe you can have a look when you get the chance.

Changle

Anyway, Zen enlightenment is an experience of a certain attitude, which a robot or brain-implant computer chip will have a hard time to experience as it doesn't possess consciousness. So it's not entirely clear how AI or brain chips can help with enlightenment, though they can do a lot of interesting things.

Ben

Well, if human consciousness is related to quantum entanglement, then maybe we could enhance human consciousness and ease human enlightenment using quantum computer brain chip implants! But I guess we're going pretty far out here, and I've taken up a lot of your time – it's probably a good point to draw the interview to a close.

One thing that comes through clearly from your comments is that the Buddhist perspective on mind is a very rich and deep one, which nudges AI and robotics research in a certain direction, different from how Western philosophy habitually nudges it. For instance, Western AI theorists are constantly arguing about the importance of embodiment for intelligence, but for you it follows immediately from nonduality. And the Buddhist theory of mind gives you a very clear perspective on which aspects of human intelligence can be expected to be achievable in AI robots. I don't agree with all your views but I think AI and robotics can use a lot of different ideas and perspectives. So I hope we can get more and more cooperation between Western and Chinese AI and robotics research, and get more and more cross-pollination and synergy between the different approaches to mind.

Anyway – thanks for sharing all your insights!

---

[184] http://multiverseaccordingtoben.blogspot.com/2009_06_22_archive.html

ACKNOWLEDGEMENT: *This interview was conducted in a complex mish-mash of English and Chinese. Copious thanks are due to Xiaojun Ding (Changle's wife, who is also a professor of Western culture and cognitive neuroscience) and Ruiting Lian (my own wife, a computational linguist) for assistance with translation, including translation of some rather subtly untranslatable Buddhist philosophical terms (some with roots in Sanskrit as well as Chinese).*

Hugo DeGaris:
Is God an Alien Mathematician?

Interview first published January 2011

Here we have another meeting of the minds between Hugo and Ben, this time focusing on Hugo's interesting quasi-theological ideas about AGI, computation and the universe.... The title summarizes the main theme fairly well!

Ben

Hugo, you've recently published an article on KurzweilAI.net titled "From Cosmism to Deism", which essentially posits a transhumanist argument that some sort of "God" exists, i.e. some sort of intelligent creator of our universe – and furthermore that this "creator" is probably some sort of mathematician. I'm curious to ask you some questions digging a little deeper into your thinking on these (fun, albeit rather far-out) issues.

Could you start out by clarifying what you mean by the two terms in the title of your article, Cosmism and Deism? (I know what I mean by Cosmism, and I described it in my Cosmist Manifesto[185] book, but I wonder if you have a slightly different meaning.)

Hugo

I defined these two terms rather succinctly in my kurzweilai.net essay, so I'll just quote those definitions here. Deism is "the belief that there is a "deity" i.e. a *creator* of the universe, a grand designer, a cosmic architect, that conceived and built our universe." Cosmism is the "ideology in favor of humanity building *artilects* this century (despite the risk that advanced artilects may decide to wipe out humanity as a pest). Artilects are "<u>arti</u>ficial intel<u>lects</u>, i.e. godlike massively intelligent machines, with intellectual capacities trillions of trillions of times above the

---

[185] http://cosmistmanifesto.blogspot.com/

human level." *Deism* is to be distinguished from *theism*, which is the belief in a deity that *also* cares about the welfare of individual humans.

Ben

Previously you have talked about "Building Gods[186]" as the ultimate goal of artificial intelligence technology. So is your basic argument in favor of deism that, if we humans will be able to build a god once our technology is a bit better – then maybe some other intelligence that came before us also was able to build gods, and already did it? And maybe this other intelligence built (among other things) us?

Hugo

Yes, pretty much. The traditional arguments in favor of a deity (as distinct from a theity, which I find ridiculous, given that last century 200-300 million people were killed in the bloodiest century in history – so much for a *loving* deity) are less strong and persuasive in my view than the artilect-as-deity argument. The rise of the artilect is based on science, and the extrapolation of artilectual intelligence to trillions of trillions of times above the human level, seems very plausible this century. If human beings (e.g. Prof Guth at MIT) have theories on how to build baby universes, then perhaps artilects could actually build them, and hence, by definition, become deities (i.e. creator gods). That was the main point of the Kurzweil ai.net essay.

Ben

I see...This is actually very similar to the "simulation argument[187]" made by Nick Bostrom, Stephen Wolfram, Jonathan vos Post and others – that since creating computer simulations as complex as our universe is probably possible using advanced technology, the odds seem fairly high we're actually living in a

---

[186] http://video.google.com/videoplay?docid=1079797626827646234

[187] http://www.simulation-argument.com/

simulation created by some other intelligences. But Bostrom, for instance, focuses on the simulation itself – whereas you seem to focus on the intelligent entity involved, the assumed creator of the simulation.

And this brings up the question of what intelligence means. What's your working definition of "intelligence"? In other words, how do you define "intelligence" in a way that applies both to human minds and to the potential super intelligent universe simulation creating deity that you hypothesize?

Hugo

As a common sense, man-in-the-street definition, I would say, "Intelligence is the ability to solve problems quickly and well." As a research scientist, I am made constantly aware, on a daily basis, of the fact that intelligence levels differ greatly between individuals. I spend my time studying PhD level pure math and mathematical physics, trying to wrap my head around the works of Fields Medal winners such as Ed Witten, Michael Freedman, Richard Borcherds, etc., all mostly likely with extraordinarily high IQs. Believe me, with my only moderately high intelligence level, it gives me "brain strain". So it's easy for me to imagine an ultra-intelligent machine. I only have to imagine a machine a little bit smarter than these genii. I am in awe at what these genii create, at what the best examples of the human intellect are capable of. I am in awe. However at a neuro-scientific level, we don't know yet what intelligence is. A five year old can ask questions about the nature of human intelligence that are beyond state-of-the-art neuroscience to answer, e.g. "What was so special about Einstein's brain that made him Einstein?" "Why are some people smarter than most?" "Why is the human brain so much smarter than the mouse brain?" I dream of the creation this century of what I label "Intelligence Theory (IT)", that would provide real answers and understanding to such questions.

We should aim at a universal definition of intelligence that would be applicable to all levels of (humanly) known intelligence. It is an interesting question how far up the superhuman intelligence level a human concocted IT could go. One would think that the finite level of human intelligence, by definition would preclude humans thinking of an IT at a level that an artilect could manage.

Ben

Following up on that, one question I have is: If there is some "superintelligence" that created the universe, how similar do you think this superintelligence is to human intelligences? Does it have a self? a personality? Does it have consciousness in the same sense that humans do? Does it have goals, make plans, remember the past, forecast the future? How can we relate to it? What can we know about it? *Hugo:* My immediate reaction to that question is that with our puny human brains, we very probably can't even begin to conceive of what an artilect might think about or be like. If we think that a universe-creating, "godlike" artilect has the human like attributes you list above, then that might a "category mistake" similar to a dog thinking that human beings are so much smarter and capable than dogs, that they must have many more bones lying around than dogs do. One thing that is interesting about this question though, is that by conceiving of the artilect as a scientific based creation, we can begin to attempt answers to such questions from a scientific perspective, not a theological one, where theologians are all too likely to give all kinds of untested answers to their particular conception of god. Is a consciousness, or sense of self a prerequisite to the creation of superhuman intelligence? These are interesting questions, that I don't have answers to. Perhaps I haven't thought deeply enough about these types of questions.

Ben

In a nutshell, how does your deism differ from conventional religions like Christianity, Islam, Judaism and so forth? And how does it differ from Buddhism, which some have argued isn't really a religion, more of a wisdom tradition or a practical philosophy?

Hugo

Traditional religions such as the above, that were invented several thousand years ago, after the agricultural revolution and the growth of cities (with their occasional genius level priest-theologian) I find ridiculous, utterly in conflict with modern scientific knowledge. The cultural anthropologists of religion have shown that humanity has invented on the order of about 100,000 different gods over the broad sweep of history, and across the planet. These many gods are so obviously invented (e.g. New Guinea gods have many pigs, etc.) that their human origins are obvious. However, the fact that every primitive little tribe has invented its own gods makes one deeply suspicious that human religiosity is in fact physiologically based, and hence has Darwinian survival value (e.g. if you can believe in a loving god, you are less likely to commit suicide in a cold, indifferent, callous universe, so religiosity inducing genes would be more selected for).

Deism, on the other hand, especially with the artilect-as-deity argument, is much closer to modern science in its conception. The very real possibility of the rise of the artilect this century virtually forces anyone confronted with the argument to accept its plausibility. Our sun is only a third the age of our universe, and there are a trillion trillion 2nd generation stars that we can "observe" that probably have generated life and intelligence. Once a biological species reaches an intelligence level that allows it to have mathematics and science, it is then only a small step for it to "move on" to the artilectual stage, whose potential intelligence is astronomically larger (pun intended) than any biological level. An artilect of the distant past in an earlier universe may have designed and built our universe. It would have been our deity.

Ben

Traditional religions serve to give people comfort and meaning in their lives. Do you think that the form of deism you advocate can serve the same purpose? Does it serve that purpose for you?

Does it make you feel more meaning in your existence, or in existence in general?

Hugo

I look down on traditional "religionists" as ignorant deluded fools. The fact that where I lived when I was living in the US, namely Logan, Utah, there were hard-science professors who were converted Mormons, showed me that the human brain is modular, with some compartments isolated from other parts, e.g. the religious areas from the critical analytical scientific areas, so that these professors were unable or unwilling to destroy their religious beliefs with just a little analytical scientific thinking. I don't have much patience with people who have low "RQs" (reality quotients). If I present these religionists with the idea that many tens of millions of theists last century were killed in the bloodiest century in history, they just block thinking about its implications. If I show them the evidence that humanity has invented 100,000 gods, they do the same. I don't deny that if one is able to believe in a loving god, it might be comforting, especially to someone who is, in Darwinian terms, sexually unattractive, and gets no human partner, so remains unloved, especially older widows, whose supply of men has run out due to the greatest of human realities, death. But emotional comfort and high RQ may not be compatible. If forced to choose, I prefer not to be a gullible child. A non-theist deist looks at the massively indifferent universe as a given. Having "faith" is no argument to me. Believing something simply because one wants to believe it allows one to believe in the "tooth fairy."

Accepting the idea that a deity might be possible, certainly increases my sense of awe. Imagine (if that is humanly possible) the capabilities of such a creature that can design and build a whole universe. That is why I call artilects "godlike". They may have godlike capabilities, but still can be thought about (by humans) as falling within the domain of science. Such a possibility makes me look on existence with a different light. I would then see the universe as having a meaning, i.e. the meaning given to it by its artilect creator. Of course, one can then ask, how was the artilect that created our universe itself created? The ultimate causation question, simply gets put back a step. The ultimate existential question "Where did all these embedded universes come from and why?" remains as mysterious as ever. But, thinking about what was going on in the "head" of an artilect deity when it designed our universe (with all its wonderful mathematical physical design) is fascinating to a scientist. How to design a universe? What a wonderful challenge for science to grapple this century and beyond. Of course, as humans, we may be too stupid to answer such a fascinating question.

Ben

I'm curious what is your own history with religion. Were your parents religious; were you brought up in a religious environment at all? I know you lived for a while in Utah, a very religious part of the US, and found that a bit uncomfortable.

Hugo

My parents were Church of England and sent their 3 kids to private Methodist schools. So until my teens I was conventionally religious, having to listen to "Christist prayers" every morning at "school assembly". I was rather late going through puberty, so my brain didn't start becoming adult and critical until I was 17. I then "discovered" science with a vengeance, and decided that I would not become a doctor but a scientist. Once I absorbed the basic scientific credo of "test your hypotheses", my old religious beliefs began to look more and more ridiculous. I then met an atheist who was a few years older than I was and very smart. What he was, served as a model for me, as to what I could become, so I rapidly switched to non-theist beliefs. The more science I learned the more ignorant traditional, 2000 year old Christist beliefs appeared to me. For decades I was an "unquestioning atheist", until the "anthropic principle" came along in the 1980s (i.e. the values of the constants in the laws of physics are so *fantastically* finely tuned to allow the existence of matter and life, that it looks as though the universe was designed) and the more math physics I learned, the more suspicious I became that the universe was designed according to highly mathematical principles – the deity-as-mathematician argument. These two principles – the "anthropic principle" and the "mathematical principle" feature in the kurzweilai.net essay *(link here)*.

Ben

Now I'm going to get a little more technical on you. You've spoken of the "deity as mathematician" argument. Is this a version of Eugene Wigner's observation of the "unreasonable effectiveness of mathematics"? It seems to me that this is an interesting intuitive argument for the existence of some fundamental hidden order in the universe – related to the order we see in mathematics – but not necessarily a strong argument for an actively intelligent "deity" with its own coherent memory, consciousness, goals, and so forth. Can you explain how the observation of surprising amounts of mathematical structure in the universe suggests the existence of a "deity" rather than just a "subtle hidden order"? Or is your deity basically the same thing as what I'm (somewhat awkwardly) calling a "subtle hidden order"? Hopefully you can see what I'm getting at here; unfortunately English isn't really ideal for discussing such things with precision (but if I switched to Lojban I'd lose most of the audience, including you!).

Hugo

Yes, subtle question. I think the rise of the artilect with its massive intelligence levels this and later centuries makes very plausible that our universe operates according to such deep mathematical principles. These principles would be the result of the artilect deity's design. Whether such principles could "be there" without such design, is hard to imagine. The deeper the physics genii of this century (such as Ed Witten, etc.) delve into the deep structure of our universe, the more mathematical it seems to be, e.g. with superstring theory using the very latest ideas in low dimensional topology, with its beautiful mathematics. This creates in my mind the deep suspicion that our universe is designed according to such mathematical principles. If it is not designed, then is it just pure chance that our universe is so highly mathematical? That seems so implausible. This "mathematical principle" is closely analogous to the "anthropic principle" in the sense that our particular universe design seems so fantastically a priori improbable. One is virtually forced to accept it has been designed. The so called "designer" traditionally was conceived of as a theity, but now that we humans can image artilects, we have a new way to imagine the designer, i.e. as an artilect, and hence compatible with our deeply held scientific principles. I guess what I'm saying is – "artilectual deism is compatible with science", whereas "traditional theism is simply pre-scientific garbage." You (may have) alluded to Spinoza's ideas with your "subtle hidden order". Einstein talked about "der Alte" (the "old one", who designed the universe). He wanted "to know his thoughts."

I agree with you that if there were no artilect-deity concept, then the existence of a subtle hidden order would support the idea of a creator less strongly. But science based artilects are now very credible, so give strong support to the idea of our universe being designed by an earlier artilect in a previous universe. One fascinating question this raises in my mind is the status depth of mathematics. Are the principles of mathematics in some sense "deeper" than even the artilect deities? Are such artilects obliged to use mathematical principles as a given, or are these principles, in some (humanly unfathomable?) sense, concocted by these artilects? This is a really deep mystery for me, but fascinating philosophically.

Ben

Hmmm... You say "If it is not designed, then is it just pure chance that our universe is so highly mathematical?"

But it seems to me that an alternate hypothesis would be *self-organization*... That is: Perhaps abstract mathematical structures are in some sense "attractors" of hyper-physical self-organizing processes. Imagine some primordial "stuff" self-organizing into a more concrete state. At first it's so abstract it's not governed by any specific physical laws, but then physical law gradually emerges within it. This is the great physicist John Archibald Wheeler's notion of "law without law[188]" Then imagine that this self-organization process inexorably tends to lead to the formation of physics-es with the same habitual deep structures. In vaguely the same way that multiple molecules with different structures tend to form similar crystalline patterns. Because these particular patterns are attractors of crystalline self-organization.

So, it could be that math is what tends to self-organize when primordial stuff concretizes into universes with concrete laws of physics. That's pretty much what John Wheeler thought.

Of course, that "law without law" idea doesn't contradict the idea of a deity that was constructed by prior advanced civilizations. But it's an alternate possible sort of explanation for why there might be so much abstract math in the universe, and it's not a boring facile explanation like "random chance".

Anyway... While we're on the topic of physics, I'm also intrigued by your notion of hyper-physics – i.e. the study of the physical laws of all possible universes, not just the one we happen to live in. But I'm perplexed by the role played in this notion by the choice of one's mathematical axiom system. It seems to me that if one has a fixed choice of mathematical axiom system (say, standard Zermelo-Frankel set theory, or whatever), then one can ask which "physical law sets" (which "physics-es") are consistent with this axiom system. So for instance, if one has a theory of what kinds of sets qualify as "space-time continua", one can then ask what kinds of space-time continua are possible according to ZF set theory. But then the question becomes: Where does the axiom system come from? Godel showed us that there's no one correct choice of mathematical axiom system. So it seems to me that hyperphysics ultimately rests on an "arbitrary" choice of

---

[188] http://what-buddha-said.net/library/pdfs/wheeler_law_without_law.pdf

mathematical axiom system, if you see what I mean. You can't get away from making some kind of foundational assumption, if you want to say *anything*. Or am I somehow misunderstanding your idea? Do you think there's some special distinguished mathematical axiom system governing all the universes in the hyperverse? If so, which one is it? Or maybe this is something only the transhuman mathematical deity knows?

Hugo

Your questions are getting deeper and subtler. I had to think about this question a while to get its essence (maybe). I interpret your question to mean "How to map the hyper-physics to a mathematical axiom system?" The ZF system currently used by us seems to work for our universe. Our (human) mathematics seems to be sufficient to understand the physics of our universe. Whether it may prove sufficient for a hyper-physics is a deep and unanswered (unanswerable?) question. As humans, it is possible that we may never get an answer to that question. Our human intelligence level is finite. There are probably many deep questions that we are too stupid to find answers to. There may be many other questions too subtle for us as human beings to even conceive of. Just how deep does mathematics go? Are we humans evolved to be mathematical? Perhaps the universe was built according to mathematical principles, hence for Darwinian survival reasons, our brains were forced to evolve to think in mathematical terms to interpret the behavior of our universe, to survive.

Ben

Yeah. The point is that the scope of all "possible" physics-es is (explicitly or implicitly) delimited by some mathematical axiom system. So it seems to me that what happens is

a) a mathematical axiom system is, somehow, chosen

b) within that axiom system, many physics-es are mathematically possible, each corresponding in principle to a different universe

So one question, in your transhumanist mathematical theology, would be whether deities can choose not only to build different possible universes, but also to operate within different mathematical axiom systems, each giving a different definition of what universes are possible.

In a sense, a mathematical axiom system serves as a "meta-physics", right?

Arguably, if a deity operates internally according to a certain mathematical axiom system, then it can't create universes outside the scope of this axiom system, except by making choices that are "random" from its perspective. So mathematical axiomatizations might form a constraint even on deities, to the extent these deities are rational and logical. On the other hand, deities might totally transcend our petty human concepts of logics, axioms and laws.

So all in all, I guess the basic question I was asking is whether you think mathematical axioms serve as a constraint on deities, or whether deities can create their own mathematical systems and structure universe based on them freely.

But of course, none of us stupid little humans knows the answers to those questions, do we? So asking such questions is more useful for exploring the scope of possibilities, than for getting at answers in the current timeframe. Although, if you do have an answer, that would be great

Hugo
When you put it like that, it's very clear, I have no problem getting your question. It is a fascinating one, and I have as yet no thoughts on the matter. Your question stimulates me though.

The only thing I've come across related to it, was the SF novel by Sagan "Contact" in which at the very end, the hyper intelligence that kept an inventory of life forms in the galaxy was smart enough to put a message in the decimal digits of the value of pi. You may remember that lovely little twist at the very end of the novel.

My human intuition is that math is prior to the deity. That even deities must obey them.

Ben
Hmmm, OK. My intuition is otherwise: I tend to think a mathematical axiom system is a representation of a certain point of view, rather than something absolute in any sense. But I don't see how to resolve the difference of intuition at this stage... We're really reaching far beyond the scope of anything either of us could be confident about.

Hugo
John Casti (of the Santa Fe Institute) says that people go thru "religious experiences" when Godel's Theorem really clicks in their minds once they have really studied him. I'm wondering if you have. Did you have a similar experience? Did you feel the full impact of his great work? I haven't yet, but would like to. It's on my list of things to do.

[EDITOR's NOTE: Kurt Godel's celebrated Incompleteness Theorem, proved in the 1930s shows that no formal mathematical system powerful enough to handle reasoning about basic algebra, can be both complete and consistent. So it shows that no finite formalization can encompass all of mathematics. This is the most important result in the area of "meta-mathematics." Hofstadter's book "Godel, Escher, Bach" is a classic, non-technical exposition of the idea and its various conceptual implications.]

Ben
Chaitin's proof of Godel's Theorem is the one that I could relate to best; when that clicked it was an "aha" moment, indeed): "You can't prove a 20 pound theorem with a 10 pound formal system"! The notion that math goes beyond any particular axiom system is obvious once you've internalized it, yet philosophically quite

deep. It's what underlay one of the questions I asked you in that interview on Cosmism and Deism... I asked whether you think there is one specific axiom system that all of the universes in the hyper-universe must obey. You replied that "even AGI gods must obey mathematics." But what does this really mean? Obey what mathematics? Obey what axiom system, if no axiom system really encompasses all mathematics, and each one (including ZFC for example) is just a partial and subjective view? Maybe each AGI god mind views math using its own axiom systems? But then how do you choose between axiom systems? Within a specific physical universe, you can say that some axiom systems are better than others because they're more convenient for modeling physics. But when you start talking about AGI gods that can build their own physical universes, then you can't say their choice of axiom systems is constrained by physics. Rather, the physics they choose will be constrained by their choice of axiom system; and which axiom system to choose becomes a matter of their "divine aesthetic taste"; a subjective matter.

Hugo

It's becoming clearer to me that I did not understand the full impact of your question on the Deism essay, regarding which axiom system the deity would use. I will study Godel a lot more, and computer science theory as well (computability etc.).

Ben

So let's get back to the practical a bit, to the notion of mathematical deism as a *religion*. Right now your approach to deism is unusual, whereas religions like Christianity, Islam, Hinduism and Buddhism occupy most of the world's population. What do you think are the prospects for a broader acceptance of your form of mathematical deism? Do you think this will become a common religion among human beings as the Singularity gets nearer? Do you think it can help people deal better with the unsettling nature of the technology-driven transformations to come? Perhaps making Cosmism more appealing to people who are attached to some sort of religious point of view? Do you envision the emergence of actual Mathematical Deist churches, where people sit and ritually collectively worship mathematical order, and the priest recites great theorems to the crowd? Where is this all going?

Hugo

I spent about 30 years of my life living in western Europe, which is highly secular. Traditional religions have pretty much died out, especially in countries like the UK, Scandinavia, etc. People are becoming better informed about the basic principles of science, so will be more accepting of a science-based deism. But, since this is a rather intellectual conception, it may only be popular with the "sages" (my word for the intellectual, high IQ types – I intend writing a book in the future, on Sagism, which hopefully will raise people's consciousness that sages are discriminated against in the modern world). As the "species dominance debate" (i.e. should humanity build artilects this century or not?") heats up in the next few decades, the Cosmists (i.e. people who want to build artilects) will use the "building gods" argument as one of their strongest, to persuade people to choose to build artilects. As secularism becomes more widespread, as theism dies, then the Darwinian religiosity components of our brain can then be "satisfied" with a "science based religion", i.e. Cosmism, the ideology in favor of building artilects. I see the "religiosity argument" of the Cosmists, being their strongest. Will there be Cosmist churches? Maybe – for the less intelligent. Churches are for the masses. Cathedrals evoke a religious response. They are bigger than a human being, making people think about higher things than where their next meal will come from. Maybe some reader of this essay will start the "Globan Cosmist Church." (Globa is the name I give to a global state, the topic of my second book "Multis and Monos" (amazon.com)). I've seen a video on YouTube of some Scandinavian guy invoking the name of Turing as a future god, with "religious gestures" and incantations using words such as the bit, the byte, etc. It was quite hypnotic. I felt like rushing out into the street shouting, "I've been forgiven of my bugs, saved by the great compiler in the sky."

But how about you? What's your view on deities and math? Have you given the question much thought?

Ben

Hmmm... I do feel there's something deeply spooky in the universe, related to the omnipresence of subtle math structures in the universe. And I've had my share of strange spiritual experiences, which have made me sometimes feel very directly in contact with transhuman intelligences – not necessarily (though maybe) creators of universes, but intelligences that are very powerful and certainly not human. So, at a gut level, I do feel there are intelligences greater than ours out there somehow, some "where." But, whether these intelligences created our universe or somehow co-evolved with it via some self-organizing process. On that I really have no strong intuition.

Really, I find that the older I get and the more I think about the Singularity and the real possibility of transhuman intelligence, the more humbled I am regarding the tremendous ignorance of the human race, and our limited capability to understand the universe. Do these words of ours like "deity", "mathematics" and "create" really say anything much about the universe, or are they pretty much as limited as a dog's conceptual vocabulary, compared to how a post-Singularity mind is going to understand things?

Hugo

Gosh, I didn't know you had had experiences with higher powers. Do you take them seriously, or interpret them as happening while you were in a certain state?

Ben

Heh... I suppose I take them about as seriously as I take everything else – which is to say, not totally! I go through my life intensely aware that Hume's problem of induction is not solved, i.e., we have no sound logical or empirical reason to believe the sun is going to rise tomorrow just because it's risen on so many previous days. Every time you pick your foot up and then set it down again assuming the ground is still going to be there, you're making a leap of faith! So I suppose I interpret my peculiar spiritual experiences about the same as I interpret all my other experiences. If I find out there are superintelligences out there I won't be shocked, and nor will I be shocked if I find out this world we're living in is just some video game created by some bug-eyed alien in his kindergarten programming class (and that he got a bad grade for his project because the humans in his creation are so extremely stupid ;P). Nor will I be shocked if it turns out the modern scientific consensus view is correct and we're the only intelligent life-forms in a physical universe that was created by nobody for no purpose whatsoever.

The main point, to me, is to improve our intelligence and our wisdom and our scope so that we can understand more and more of the universe and ourselves. I don't know if there's a deity or if our universe is a simulation, nor if the deity is a mathematician if there is one, but I'm sure I'll understand these ideas better if I can multiply my intelligence by a factor of 100. If I'll even be "me" at that point, anyways... But that's another thorny futuristico-philosophical question, better left for another interview!

Lincoln Cannon:
The Most Transhumanist Religion?

Interview first published May 2011

According to my informal observations, the majority of transhumanists don't consider themselves affiliated with any traditional religious organizations or belief systems – though some, like Giulio Prisco[189], are deeply interested in the creation of new spiritual traditions founded on transhumanist ideas. However, there is also a nontrivial minority of transhumanists who combine their transhumanism with traditional religious beliefs and membership in traditional religious organizations. And among the most vocal of the religious transhumanists has been the Mormon Transhumanist Association[190] (MTA) – a group consisting of of 116 members, with approximately 41% living in Utah and 90% living in the United States.

In 2011 Hank Hyena interviewed Lincoln Cannon[191], the co-founder, director and president of the MTA, for H+ Magazine. My goal in doing this follow-up dialogue with Lincoln was to dig a bit deeper into the related issues, and try to understand more fully how Lincoln and his colleagues bring together the two thought-systems of Mormonism and transhumanism, which at first glance would seem very different.

Ben
Since I like to get right to the heart of things, I'll start off with a fairly blunt question. As you already know, I'm not a religious guy. I'm open to the validity of individual and collective spiritual

---

[189] http://hplusmagazine.com/2011/02/08/technological-transcendence-an-interview-with-giulio-prisco/
[190] http://transfigurism.org/
[191] http://hplusmagazine.com/2010/08/22/mormon-transhumanism-eternal-progression-towards-becoming-god/

experiences – but the belief systems of the world's major religions tend to strike me as pretty absurd... And Mormonism is no exception. Some of the basic teachings of Mormonism, as I read them, seem plainly ridiculous by the lights of modern science. That is, they describe things that seem extremely unlikely according to known scientific theories, and for which there is no available empirical evidence.

A couple random examples are the virgin birth of Jesus, and the literal existence of Celestial and Telestial Kingdoms, etc. etc. You know what I mean... You're a very well educated, scientifically and technically literate guy. How can you believe that crazy stuff??

Lincoln

From some perspectives, aspects of Mormonism are indeed absurd. To paraphrase one prominent atheist, Mormonism is just Christianity plus some other crazy stuff. However, these perspectives overlook or ignore how the other crazy stuff modifies the Christianity! It does so to such an extent that characterizing Mormonism as a mere extension of other modern Christian ideologies is inaccurate. Mormonism is to modern Christianity as ancient Christianity was to Judaism. It is a different religion.

The virgin birth is an interesting case in point. On the one hand, because Mormonism rejects the mainstream Christian notion of original sin, the virgin birth has somewhat less theological significance. All children are sinless, and none sins until capable of moral reasoning, so Mormons have no need of explaining how Jesus could be born sinless. On the other hand, the virgin birth still has some theological significance for Mormons that use it to explain how Jesus could continue to live sinlessly even after maturing to an age capable of moral reasoning. From their perspective, Jesus gained a special moral capacity because of his unique conception. Personally, I esteem Jesus as a principal model of morality simply by definition, rather than because of any special conception that enabled him to measure up to an external model of morality. As I see it, the virgin birth is primarily a reflection of ancient symbolism that would direct our human idealizations to and beyond the resolution of moral conflict.

What of the literality of the virgin birth? Is it compatible with modern science? Mormons are philosophical materialists, with scriptures teaching that all spirit is matter, and God has a material body. Accordingly, some prominent early Mormons (notably Brigham Young) speculated that God and Mary conceived Jesus through natural means. That idea bothers some people, including many modern Mormons, some of whom prefer mainstream Christian perspectives on mechanisms for the virgin birth. Others have speculated that God could have used artificial insemination. Personally, while I find the question of the literality of the virgin birth interesting, I also consider it trivial, particularly as a factor in my esteem for Jesus. Perhaps the virgin birth is exclusively symbolic; perhaps the matrix architect, so to speak, intervened. Emotionally, I'm indifferent, except that I recognize moral and practical reasons to embrace symbols while also rejecting supernatural (empirically inaccessible) explanations in all matters.

How about the varying degrees of heavenly glory described in Mormon scripture and ritual? Are they, literally interpreted, compatible with modern science? That depends on how one understands their literal interpretation, which is complicated by their overtly symbolic and esoteric descriptions. Here's an interpretation that I consider both literal and symbolic. Presently, we live in one of innumerable telestial heavens. Eventually, if all goes well, our heaven will become a terrestrial heaven (also described as a millennial world), wherein present notions of death and poverty will no longer apply. Subsequently, again if all goes well, our terrestrial heaven will become a celestial heaven, whose inhabitants become capable of creating new heavens, repeating the cycle. Each transition depends both on the context of opportunity provided by the grace of God, and on our own work to learn of and adhere to principles whose contextual consequences are increased flourishing and eventual deification, both individually and communally, as well as environmentally. Some Mormons hold that these changes are inevitable at the communal and environmental levels, whereas there is real risk only for individuals. Others, such as I, interpret scriptures that suggest inevitability as psychological motivators rather than absolute foretellings, and recognize real communal and environmental risks. All of this should sound vaguely familiar to transhumanists, most of whom hold to notions of human flourishing through various stages (perhaps articulated in terms of the Kardeshev scale), many of whom imagine we'll eventually prove capable of computing new worlds as detailed as our own (thereby implying we are almost certainly living in a computed world ourselves), and some of whom engage in disagreement over how best to consider and articulate the evitability of these changes.

There are probably several other Mormon teachings that strike you and others as being incompatible with modern science, and I'd be happy to discuss any others that interest you. More generally, though, it's worth noting that most Mormons value scientific education. Accordingly, geographical areas of higher than average Mormon concentration produce a higher than average number of scientists per capita, and among Mormons there is a positive correlation between level of education and level of religious activity. Many Mormons, such as I, consider the scientific project to reflect the basic principles of our faith, among which is the search for and acceptance of all truth from any source, to paraphrase founder Joseph Smith. Of course, that implies we have much to learn, and we should readily acknowledge that. Yet persons unfamiliar with Mormonism that hear of Mormon ideas that sound incompatible with science should not assume that educated Mormons are casually dismissing science. We probably have something at least somewhat intelligent to say about the matter.

Ben

OK, now let me come at things from a different direction. While I value science a lot, and I've never been religious, I realize science is not the only source of truth — I've often been impressed with the insights of Zen masters and other spiritual visionaries. It seems that every religion has a side that focuses more on pure spirituality than on superstitions — Buddhism has Zen; Islam has Sufism; Christianity had Meister Eckhart and so forth. Aldous Huxley's book "The Perennial Philosophy" argued that the deepest spiritual aspects of all the different religions were basically the same, and tried to summarize the commonality. Would you say there also exists a stripped-down, purely spiritual aspect to Mormonism, which espouses something reminiscent of Huxley's Perennial Philosophy?

Lincoln

Mormonism has a side reminiscent of the Perennial Philosophy, yes. Mormon scripture describes the presence of God as the light in and through all things, and states both that we are seeing God when we look into the heavens and that God enlightens our own eyes and understandings. God is both transcendent and immanent, in the neighbor we should serve and in the Earth that bemoans our immorality. Awareness of this should lead us to seek to become one with God, achieving deification not in the egotistical sense of raising ourselves above others, but rather in the altruistic sense of raising each other together.

These ideas have strong parallels in many other religions, as suggested in Huxley's Perennial Philosophy, Campbell's Monomyth and others. Many Mormons acknowledge these similarities, at least on some level. For example, Joseph Smith acknowledged commonalities with the Christian sects of his day and claimed we would not enjoy the millennial world until Christians cease arguing with each other. Modern Mormon authorities regularly comment on their observations of inspired teachings within ideologies ranging from classical Greek philosophy to Asian religion. The Book of Mormon also includes passages that indicate God speaks to everyone everywhere, without regard to religion, race or gender. Of course, Mormons haven't always internalized and acted in accordance with the universalist aspects of our faith. We have significant cases of both racism and sectarian hostility in our history. I expect, however, that we will continue to improve in this area, along with the rest of the world.

Incidentally, most Mormons do not consider the mystical aspects of our religion to contradict the idea that God is material and corporeal. While explanations for compatibility vary from Mormon to Mormon, my own speculation is that our universe is part of God, like software is part of a computer or an embryo is part of its mother. As our computational capacity has increased, shrinking in both cost and size, it has also become more intimate, moving from distant warehouses into our pockets and even our bodies. We are decreasingly distinguishable from our computers, and it seems reasonable to suppose that posthumans would be altogether indistinguishable from their computers. For such beings, there may be no practical difference between thinking of a world and creating it. We can imagine them as both materially corporeal and meaningfully present throughout the worlds they create.

Ben

Interesting. That does help me to understand Mormonism better. Now let's turn more explicitly toward transhumanism…

I wonder: Do you see Mormonism as consistent with "traditional transhumanism"? In the typical transhumanist view, to paraphrase Nietzsche, "man is something to be overcome — via technology." Humans are seen as something to be improved and purposefully adapted, perhaps indefinitely until they become something radically different than what is now conceived as human. On the other hand, it seems Mormonism presents a different visions of the future, centered on the Second Coming of Christ and the subsequent ascension of good people to Heaven and descent of bad people to Hell, etc. It's not clear to me how these two different visions of the future can be reconciled; could you clarify for me?

Lincoln

Mormonism has many parallels with traditional Transhumanism. In Mormonism, natural humanity is something to overcome as we learn to become more like God. God graciously provides means (technological and otherwise) for us to progress, and we must use these means instead of merely supposing God will save us without any effort on our part. As we become more like God, we will change both spiritually and physically, taking on the virtues and attributes of God, including both creative and benevolent capacities. As described in Mormon scripture, future physical changes will include transfiguration of the living and resurrection of the dead to immortality in material bodies, varying in glory according to the desires and works of each individual. The mainstream Christian notion of a simple dichotomy between heaven and hell is not part of Mormonism. Instead, heaven and hell are states of being, categorized into degrees of glory, such as the terrestrial and celestial glories discussed previously.

Mormonism's anticipation of the Return of Christ parallels the full range of apocalyptic and messianic expectations of many Singularitarians. According to Mormon prophets, the time in which we live is one of hastening progress toward a millennial world of paradise, but catastrophic risks attend. Some are persuaded that the prophecies, bad and good, are inevitable. Others, such as I, consider the prophecies valuable as forth-tellings of opportunities to pursue and risks to avoid, presenting practical and moral imperatives to trust that we can make a positive difference. Some subscribe to rigid interpretations of how Christ will independently triumph over the apocalyptic challenges. Others, such as I, are comfortable exploring interpretations that include our participation in Christ, following Jesus' invitation to take on the name of Christ and join in the salvific work of God. The analogous Singularitarian question is: will independent artificial intelligence help or hurt us, or will we enhance ourselves such that we can participate directly in making that decision?

Ben

Hmmmm.... Some transhumanists believe that once we use technology to expand our minds far enough, we will be able to essentially come into communication with the mind of the universe — talk to God, in a sense. Does this notion have any resonance with the Mormon notion of God? According to the Mormon perspective, could enhanced future humans have a closer mental link to God than ordinary humans?

Lincoln

Mormonism shares with mainstream Christianity the idea that eternal life is to know God. Many of us consider science, and its technological enablers, one important way that we can come to know God. If we are something like the thoughts of God then to observe the universe and to try to figure out how it works is to engage in a sort of divine psychology. Even from the perspective that God created the universe externally, to study a creation is to study its creator. As Mormon scripture points out, we're already seeing God when we observe the universe, even if we don't understand.

Ben

What about Teilhard de Chardin's notion of the noosphere, which some thinkers have connected with the concept of a technologically-constructed Global Brain. The basic idea here is that computing and communication technology is increasingly linking us all together into a global web which increasingly has its own mind and intelligence. Teilhard then saw this global web mind ultimately reaching up and fusing with the mind of God in a way that's harder for us individual humans. Does this sort of idea have any analogue in the Mormon perspective?

Lincoln

Mormonism also shares with mainstream Christianity the idea that together we are, or at least should become, the body of Christ. We should reconcile and unite with each other and with God. I suspect such ideas were the origin of Teilhard de

Chardin's suggestion that Christ has a cosmic body. I would also associate the realization of such ideas with the Return of Christ.

More unique to Mormonism is the idea that Earth, when transformed into a celestial heaven, will become something like a sea of glass and fire in which all things, past and future, will be manifest to its inhabitants. The scriptures also describe the inhabitants of the celestial heaven receiving white stones through which they'll learn of heavens of higher orders. In such ideas, it's not hard to imagine future technology at work, as did Teilhard de Chardin.

Ben

Hmmm..... Well I must say that:

Earth, when transformed into a celestial heaven, will become something like a sea of glass and fire in which all things, past and future, will be manifest to its inhabitants. The scriptures also describe the inhabitants of the celestial heaven receiving white stones through which they'll learn of heavens of higher orders.

Certainly sounds funky! And not so implausible, IF you interpret it metaphorically. However, I wonder how literally you take all that. Do you take it as a general metaphor, or as a literal foretelling of the specifics of the future? And how do other Mormons generally take it — literally or metaphorically?

Lincoln

Mormon interpretations of scripture range broadly from the highly literal to the highly symbolic; however, most Mormons do not strictly subscribe to scriptural inerrancy, infallibility or literalism. Personally, I am most concerned with interpreting scripture non-dogmatically and pragmatically, in ways that are inspiring and helpful to the best of my ability to judge rationally and emotionally.

Accordingly, I don't insist on literal interpretations of scriptural descriptions of the heavens. Indeed, Mormon prophets and the scriptures themselves encourage loose interpretations. For example, one of the more lengthy scriptural descriptions of the heavens includes this revelatory question, "Unto what shall I liken these kingdoms, that ye may understand?" Implicit in this question is the indication that we would not understand a more literal description. Similarly, Joseph Smith once commented that a person would learn more about heaven by observing it for five minutes than by reading everything ever written on the subject.

Ben

One of the more interesting and radical ideas to emerge from the transhumanist perspective is the "simulation argument" (as articulated for example by Nick Bostrom), which argues that it's fairly likely our universe is actually part of a simulation purposefully created by alien intelligences in a civilization evolved previously to ours. What's your reaction to this? Does it contradict Mormon teachings or can it somehow be made consistent with them? If so, how?

Lincoln

Mormon theology, as articulated by persons that Mormons typically recognize as prophets, resonates strongly with the Simulation Argument. Joseph Smith proclaimed that God was once as we are now, became exalted, and instituted laws whereby others could learn how to be gods, the same as all gods have done before. Wilford Woodruff taught that God is progressing in knowledge and power without end, and it is just so with us. Lorenzo Snow prophesied that children now at play making mud worlds will progress in knowledge and power over nature to organize worlds as gods. Accordingly, many Mormons, such as I, have faith in a natural God that became God through natural means, suggesting how we might do the same.

The New God Argument, which I formulated with Joseph West, leverages a generalization of the Simulation Argument, as well as the Great Filter argument and some other observations stemming from contemporary science and technological trends, to prove that if we trust in our own posthuman potential then we should also trust that posthumans more benevolent than us created our world. Because such posthumans may qualify as God in Mormonism, the argument suggests that trust in our posthuman potential should lead to faith in a particular kind of God.

It's worth noting that Mormonism, because of its relatively unique theology of a progressing posthuman God, is often the target of hubris charges similar to those aimed at Transhumanism by adherents of religions with mainstream theologies. They consider Mormonism, like Transhumanism, to be committing the sin of Babel, as portrayed in the Bible. Some Mormon authorities have responded that the sin of Babel is not merely in the desire or attempt to become like God, but rather the sin is in allowing our technical achievements to outpace moral achievements, pursuing the desire foolishly or egotistically.

Ben

I'm not familiar with the New God argument, so I guess most of our readers won't be either! Could you perhaps give a brief summary?

I understand the argument that, if we have posthuman potential, then probably some other race came before us in the history of the universe and also had posthuman potential, and probably that other race created virtual worlds of great complexity, and probably we live in one of those virtual worlds. This is basically the Simulation Argument, and Hugo DeGaris combines it with the observation of surprising mathematical structures in the world to conclude that "God may be an alien mathematician."

But where do you get the conclusion that the posthumans who created the virtual world we live in, must be MORE beneficial than us? I don't get that leap; could you clarify?

Lincoln

Here's a summary of the New God Argument:

If we will not go extinct before becoming posthumans then, given assumptions consistent with contemporary science and technological trends, posthumans probably already exist that are more benevolent than us and that created our world. If prehumans are probable then posthumans probably already exist. If posthumans probably increased faster in destructive than defensive capacity then posthumans probably are more benevolent than us. If posthumans probably create many worlds like those in their past then posthumans probably created our world. The only alternative is that we probably will go extinct before becoming posthumans.

Among Transhumanists, the most controversial part of the argument is that to some, it's not clear that benevolence is the only possible explanation for surviving an increasing gap between destructive and defensive capacity. However, I think there is a strong case to make for this assumption, and I'm working on articulating it for future publication. In brief, though, the idea is that warfare selects for communal complexity, which in turn selects for individual restraint from infringing on others' interests.

Ben

Hmmmm.. yeah, as you may surmise, I don't really buy the benevolence argument. It seems to imply that we humans are among the LEAST benevolent creatures that can possibly become posthumans. But I don't see a basis for this – i.e., I don't see how we could possibly know this!.... Also, I'm not even clear that a concept like "benevolence" has to apply to posthumans, which may have an entirely differently way of thinking and acting than we can understand…

Lincoln

Your criticism of the Benevolence Argument seems insightful, judging from what I've been reading to help me better justify it.

You suggest that the argument seems to imply that humans are among the LEAST benevolent creatures that can possibly become posthumans. This observation may be a step in the direction that some anthropologists have been going. As I mentioned before, there is a theory that warfare selects for communal complexity, which in turn selects for individual restraint from infringing on others' interests. Taken to its logical conclusion, the theory implies that at any particular stage in human development, we have been both the LEAST and the MOST benevolent creatures that could possibly become posthumans (assuming we go on to become them). In other words, the consequence of highly violent pressures is the natural selection of cooperative groups that can counteract those pressures. Paradoxically, morality emerged from its opposite. This theory falls in line well with Pinker's observations about humanity's relatively violent past and relatively peaceful present.

In all of this, note that I'm using words like "benevolence", "morality" and "cooperation" as synonyms. I don't have any strict or narrow definition of benevolence in mind. Quite to the contrary, I'm willing to learn something about the nature of benevolence as I come to understand better what it is in our nature that is bridging an increasing gap between our destructive and defensive capacities. Even if not exactly as I'm imagining it, it seems something approximating what we'd call "benevolence" must be essential to the survival of posthumans, unless some kind of extremely dystopian scenario of monolithic volition is possible to maintain. Anyway, there's more work to do here, for sure.

Ben

I see, so you're saying that morality as a social and cultural pattern evolved because of the individual tendency to violence that humans have. Less violent creatures wouldn't have needed to develop so much morality.

But I still don't see why this implies that humans would necessarily be less benevolent than the previous intelligences who created the world we live in.

Perhaps a better argument that the creators of our world were probably more benevolent than current humans is as follows:

1. Right now, current humans are not all that benevolent, because we're largely controlled by our unconscious minds, which evolved for survival in a Paleolithic regime where violence was necessary.

2. However, as humans achieve more advanced technology, we will figure out how to make ourselves more and more benevolent, via education and changes in culture, but also due to brain modifications, brain-computer hybridization and so forth. We will do this because now that the Paleolithic need to fight to survive and reproduce is gone, our old violent ways have become counterproductive.

3. So, by the time we figure out how to create massive, complex simulated universes (comparable to the one we live in), we will probably be more benevolent than we are now (as we will have re-factored ourselves for greater benevolence).

4. Other prior intelligent species probably underwent similar dynamics before they became able to create complex simulated universes (like ours).

5. Thus, our world was probably created by a previously-evolved intelligent species, with more benevolence than we currently have.

Or to borrow Hugo DeGaris's vernacular, not only is God probably an alien mathematician[192], he's probably a fairly benevolent alien mathematician...

I'm not sure if the above is a new argument for the benevolence of the Creator, or if it's just a rephrasing of your argument in my own vernacular (because I haven't really understood your argument yet, the way you've worded it).

Lincoln

Yes. I think something like that approximates a justification for the assumptions on which the Benevolence Argument is based. Although humans presently aren't as benevolent as we could hope, there's good evidence that we're more benevolent (less inclined to violence and more inclined to cooperation) than our distant ancestors, perhaps consequent to adaptations that enabled us to survive our violent history. There's also good reason to suppose, as you point out, that advancing technology (both machine and social) will facilitate further increases in benevolence. However, advancing tech will also facilitate increases in capacity for violence, including global catastrophic risks that we may not survive. If we do survive, I expect the reason will be the same as it has been historically: adaptation to decreased violence and increased cooperation. This underscores the importance of transhumanism, understood not just as the arbitrary use of technology to extend our abilities, but rather as the ethical use of technology.

Ben

What are the main ways in which the Mormon spin on transhumanism differs from "conventional" transhumanism (bearing in mind that the latter is a rather diverse entity, of course...)

---

[192] http://hplusmagazine.com/2011/01/18/is-god-an-alien-mathematician/

Lincoln

Mormon Transhumanism doesn't differ from conventional Transhumanism in essentials so much as it extends conventional Transhumanism. Not content with describing our future in merely secular terms, Mormon Transhumanists embrace a religious esthetic for varying reasons. I do so because I consider the religious esthetic more powerful as a motivator and more accurate as a descriptor. Of course, divine demanders can be abused, and God-colored spectacles can distract. However, I prefer these risks to those of alternatives available to me.

Ben

Hmmm... It's obvious from history that the religious esthetic is an awesome, perhaps unparalleled, motivator for human beings. The only things that compete with it seem to be biological necessities like food and sex and protecting one's children. It's less clear to me why you consider it "more accurate". Can you give me some examples of how the religious esthetic is more accurate than other sorts of esthetics, let's say for example than a postmodern esthetic, or a Nietzschean esthetic, or a secular humanist esthetic, or a Cosmist esthetic, etc. etc.?

Lincoln

I consider the religious esthetic (in contrast to religious epistemics, for example) to be more accurate than alternatives as a descriptor of my experience. I don't experience the world in terms of reductionism, no matter how epistemically accurate reductionism may be. I do experience the world in terms of joys and sorrows, goods and evils, truths and errors, all ranging in magnitude from the mundane to the sublime. Such experience is also presented, of course, in non-religious esthetics, such as postmodernism and secular humanism. However, I find that the non-religious articulations resonate with me more as they approach religious forms. For example, I thoroughly enjoy Nietzsche, who vies with William James for status as my favorite philosopher; and my favorite work from Nietzsche is "Thus Spake Zarathustra", which clearly takes on a religious esthetic. Regarding Cosmism, I'd suggest that its accuracy as a descriptor of my experience, too, is in its approaches to the religious esthetic.

I embrace the religious esthetic of Mormonism in particular both because it's my cultural heritage and because I perceive relatively unique value in its doctrine of communal deification through the Gospel of Christ. It informs my motivations in ways that I think make me a better person, a better friend and father, and a better contributor to our common pursuit of a better world.

Ben

Could you clarify the "unique value" aspect? Communal deification is not unique to Mormonism. And what is the unique value offered by the "through the Gospel of Christ" aspect of the communal deification? In what ways is communal deification through the Gospel of Christ uniquely valuable, as opposed to other sorts of communal deification?

I guess another way to put the question is: For individuals NOT coming from the Mormon cultural heritage, but liking the general idea of communal deification, what's the reason for them to embrace the Mormon view? What does it add to "plain vanilla communal deification"?

Lincoln

The doctrine of deification is not absolutely unique to Mormonism, but it is relatively unique. The idea is ancient, but I know of no other major modern religion that openly teaches it. Even compared to most of the small religious movements that teach and have taught it, Mormonism's formulation is more concrete. For Mormons, deification is not only a mystical union with an abstract God. Rather, deification is understood as both spiritual and physical progress to material corporeal godhood.

Ben

The point that "For Mormons, deification is not only a mystical union with an abstract God. Rather, deification is understood as both spiritual and physical progress to material corporeal godhood" is certainly interesting and does distinguish Mormonism from most other religions, bringing it closer to transhumanism in some ways.

Lincoln

The combination of deification and the Gospel of Christ is particularly attractive to me because of how the Gospel of Christ informs our understanding of deification. The writings of Paul in the Bible allude to a dichotomy of would-be Gods. On the one hand is Satan, described as seeking to raise himself above all else that is called God, declaring himself God. On the other hand is Christ, described as seeking to raise us together as joint-heirs in the glory of God. These archetypes of egotistical and altruistic godhood serve as reminders that deification in itself is not the goal, but rather the goal is benevolent deification. This was echoed strongly by Joseph Smith, who taught that our salvation depends on each other, both the living and the dead, and that none can be perfected without the other.

Of course, Mormonism is not the only religion that teaches the value of benevolence. However, by combining it with the explicit goal of deification, it bids us look higher and work harder. It is not enough to be humanly benevolent, or even saintly benevolent. We are called to be divinely benevolent, taking on the name of Christ, following Jesus' example, and participating as saviors in the work of God to bring about human immortality and eternal life

Ben

Do you think that a Singularity a la Vinge and Kurzweil is in the cards? Not to put too fine a point on it — but: How can you consistently believe both in the Singularity and the Second Coming of Christ? Which one do you predict will come first, for example?

Lincoln

Although I acknowledge accelerating technological change, I have mixed feelings about self-identifying as a Singularitarian because I think a technological singularity could be a moral failing. Rather than conceding our predictive and directive capacities to independent self-improving artificial intelligence, I hope and expect we can also enhance our own intelligence, thereby expanding our predictive and directive capacities, and precluding experience of a technological event horizon.

Whether experienced as a technological singularity or not, how does accelerating technological change fit with trust in the Return of Christ? On one hand, the Return of Christ is when we, as a community, attain unity in benevolence toward each other. In that day, we become the body of Christ, returned to the prophesied millennial paradise. On the other hand, how would friendly artificial and enhanced intelligence manifest itself to and in us? Assuming we live in a computed universe, the structure of which constrains the optimization of advanced intelligence, would it reveal our posthuman creator?

"Beloved, now are we the sons of God, and it doth not yet appear what we shall be: but we know that, when he shall appear, we shall be like him; for we shall see him as he is." (1 John 3: 2)

Ben

I understand that's how you think about it as a Mormon. But I don't quite see why we can't attain unity in benevolence with each other without it having anything much to do with this dude named Jesus who lived 2000 years ago....

Relatedly and more broadly, I wonder what you feel the Mormon way of thinking has to offer transhumanists who were NOT brought up in the Mormon cultural tradition? Can you give us some compelling reasons why it would make sense for us to adopt Mormon transhumanism instead of plain old transhumanism or Cosmism or something else? Or to put it another way: Suppose you were a Mormon missionary trying to convert transhumanists: what would you tell them?

Lincoln

Individually, we can achieve unity in benevolence without it having much to do with Jesus of Nazareth. Some of my non-religious and Asian friends are excellent examples of this. Most Mormons anticipate that the millennial world will consist of persons from many different ideological persuasions. Communally, however, our achievements will have much to do with Jesus, in the least because no other person has had more influence on our human civilization. Clearly not everything done in his name has been good; but, Christian or not, we are part of a system in which he is a significant variable.

So far as I'm concerned, the only good reasons to add Mormonism to your Transhumanism would be those gained from experiment and experience. Mormon missionaries are fond of referencing a passage of scripture from the Book of Mormon that compares "the word" (in this case, Mormonism) to a seed, which we must plant and nourish before harvesting and tasting its fruit. In tasting the fruit, we learn the value of the seed. In other words, to know the value of Mormonism, one must try it. It's more than theological arguments, scriptural exegesis, and extraordinary historical claims. Mormonism is an immersive way of life, and one that happens to be remarkable among religions in its compatibility with Transhumanism.

Ben

About needing to try Mormonism to appreciate it — yes I can see that may be true, but on the other hand the same is also true of so many other things... and you're not giving us a reason to choose to try Mormonism instead of one of the many other things that can only be understood via immersing oneself in them...

Lincoln

Here are some of the practical benefits of the Mormon lifestyle:

Mormons live longer than average, probably due to our code of health and emphasis on family relations and education.

Mormons are less likely than average to commit suicide.

Mormon communities have relatively low incarceration rates, reflecting low crime rates.

Mormon communities may be the safest places to be during large scale emergencies because many of us maintain a year's worth of food and emergency supplies.

Mormons are relatively likely to help you, as reflected by Utah's consistent ranking as one of the most charitable states in proportion to income.

Mormon communities can spend less than public welfare because of generous private donations.

Mormon communities are commonly ranked by the media to be among the best places to live.

Mormons make dependable employees, with better than average education and lower than average probability of showing up to work with a hangover or buzz.

Mormon communities have some of the highest ratios of academic aptitude to dollars spent on education.

Mormon communities produce a higher than average per capita number of scientists.

Mormons are more likely than average to have computers.

Mormon communities have lower than average teen pregnancy and abortion rates, reflecting emphasis on sexual abstinence outside marriage.

Mormons make more babies than average, which may contribute to long term economic and ideological strength.

Mormons are less likely than average to become divorced.

Ben

Hah – interesting response! That's certainly a refreshingly practical perspective....! Although, I guess the practical statistics for Ashkenazi Jews are also pretty good, and that's my own heritage, so it doesn't really give me much motivation to become a Mormon, heh....

About Mormonism's unique connection to transhumanism — I can see that it does have some unique or almost-unique aspects that tie it closely to transhumanism, and that is very interesting. However, other religions such as Buddhism tie in closely with transhumanism in other ways, that are also very powerful — for instance, Buddhism views the world as constructed by the minds in it, which is very reminiscent of the Simulation Argument and in general of the transhumanist meme of constructing one's own self and reality. So I'm not sure Mormonism is uniquely closely tied to transhumanism — but I can see that it has some deep and unique or nearly unique specific ties to transhumanism. So thanks for opening my eyes in that regard!

Lincoln

I'd like to summarize for you some of the more significant parallels between Mormonism and Transhumanism. Perhaps this will help better communicate why I consider Mormonism be the strongest example of religious Transhumanism among major contemporary religions.

Mormonism posits that we are living in the Dispensation of the Fullness of Times, when the work of God is hastening, inspiring and endowing us with unprecedented knowledge and power. This parallels the common Transhumanist position that we are experiencing accelerating technological change. So far as I know, no other major religion has a strong parallel to Transhumanism in this area. This is probably because of how recently Mormonism was founded compared to other major religions.

Mormonism expects the Dispensation of the Fullness of Times will culminate in the Return of Christ, along with attending risks and opportunities. This parallels the apocalyptic and millenarian expectations that Transhumanists commonly associate with the Technological Singularity. This parallel between Mormonism and Transhumanism is shared with other major Western religions.

Mormonism prophecies that the Return of Christ will transform the Earth into paradise during the Millennium, when there will be no poverty or death, the living will be transfigured, and the dead will be resurrected. This parallels the common Transhumanist position that the Technological Singularity may facilitate radical life extension, super abundance, and potentially even an engineered omega point. So far as I know, this parallel between Mormonism and Transhumanism is shared only to lesser extents with other major religions, which interpret transfiguration and resurrection in only spiritual terms rather than including the physical as does Mormonism.

Mormon vision culminates in a plurality of Gods, eternally progressing and creating worlds without end, both spiritually and physically. This parallels Transhumanists' common expectation that we will prove capable of engineering intelligence and worlds, and reflects the logical ramifications of the Simulation Argument. It appears to me that most other major religions don't share this parallel with Transhumanism at all, and Buddhism does only in a more abstract manner.

Mormon metaphysics shares the basic assumptions of science, including consistency, causality, uniformity, empiricism and materialism, such that even miracles, although marvelous in our eyes, do not contravene law. Likewise, Transhumanists hold to the basic metaphysical assumptions of science, while anticipating engineering marvels. I know of no other major religion that shares this parallel, particularly in the area of materialism.

Mormonism aims at nearly universal salvation, in physical and spiritual immortality and eternal life, enabled in part through genealogical and proxy work for the dead. Similarly, a relatively unique Transhumanist perspective is that we might re-engineer the dead by copying them to the future, perhaps via quantum archeology. I don't think there is another major religion that shares the expectation that proxied information work can contribute to human salvation.

Mormon scripture champions glorified physical life in this world, denigrating death as an "awful monster" and declaring "more blessed" those who wish serve God in this world indefinitely without dying. In parallel, Transhumanists commonly promote radical life extension and the conquest of death in this world. Mormonism and Transhumanism together contrast with the relatively escapist advocations of other major religions that explain immortality and heaven in dualist other-worldly terms.

Ben

Very interesting, very interesting....

Oh, finally... I have to ask you one more question, even though it's not so directly tied to transhumanism....

How does Mormon thinking handle the "problem of evil" ?

That is: Why, if there's this super-powerful, super-good God, are there so many downright horrible things in the world?

Leibniz of course answered this by saying that even God must obey logic, and we live in the best of all possible worlds. Like Voltaire, I never found this fully satisfying....

On the other hand, Buddhism basically tells us that all existence is suffering, and to avoid it we need to opt out of existence and embrace the Nothing...

I suppose this does relate to transhumanism, in that IF transhumanist technologies realize their potential for benefit, then the amount of evil in the world will dramatically decrease. Presumably you believe that God will allow this to happen — but then, why didn't He let it happen already, instead of letting so many of us suffer and die? What really is the moral justification for a benevolent God letting so many apparently innocent little girls get raped and brutally murdered for example?

I'm genuinely curious how Mormonism addresses this dilemma...

Lincoln

Mormons commonly attribute evil to a couple things. First, God relinquished a systemic agency to the world. Along with God's relinquishment of human agency, the whole world was relinquished. The allowance for human agency necessitated a permissive context, which enables not only humans' volitional evil, but also cascades into the natural evils we observe and experience. Second, God did not create the world from nothing. Rather, God organized our world and its laws within a context of previously existing matter and laws, which may constrain God in various ways.

Most Mormons don't hold that we live in the best of all possible worlds. Neither do we hold that existence is suffering to escape. Instead, our scriptures teach that we exist that we might have joy, and that we can make real differences in the degree of joy we experience as we make better or poorer choices.

So why does God not intervene to give us all a fullness of joy now, or at least faster? Mormons tend to trust that God is intervening, within the limitations that present themselves, according to an awareness and wisdom far superior to our own. Our scriptures also suggest that our eventual salvation is so valuable to God that it justifies doing just about anything short of annihilating us to help us attain it. Indeed, it justified relinquishing the world in the first place, as God wept, knowing we would suffer. So we trust in the grace of God to provide a context of opportunity within which we work out our salvation and seek to become as God, as we share in the suffering.

What about the specific evils? Why might a Mormon claim that God intervened in one case but not in another? Why would God let so many innocent little girls get raped and brutally murdered? I don't know. It's horrible. I feel a divine calling to do what I can to make the world better. I imagine, assuming my trust is not vain, that God feels an analogous calling. Maybe that's just my anthropomorphized projections, but such projections seem to be of the sort that would shape us into Gods, and we almost certainly would not be the first or only to be so shaped. Paradox. Perhaps morality must arise from a context of immorality? Perhaps Christ cannot function without Satan? Maybe there is no other way? This brings us full circle to my comments on justification of the Benevolence Argument.

Ben

So this is largely the same as Leibniz's idea that logic constrains even God... But now you're saying that both logic and PHYSICS constrain even God....

But still... Intuitively, I can't really buy your idea that "God feels an analogous calling." ...

I can sort of buy the argument that our world was fairly likely created by some previously-evolved intelligent creature who liked math and was probably less violent and nasty than current humans....

But, the idea that this Creator felt a calling to minimize our pain and suffering in this simulated world we created, and work toward our betterment — that just seems to contradict what I see in the world around me. I find it rather hard to believe that the degree of horror and suffering in the world is really necessary for our ultimate betterment, enlightenment and salvation.

I find it more likely that the superhuman intelligences who created our world didn't so much give a crap about the particulars of the suffering or pleasure of the creatures living within it. Maybe they feel more like I did when I had an ant farm as a kid... I looked at the ants in the ant farm, and if one ant started tearing another to pieces, I sometimes felt a little sorry for the victim, but I didn't always intervene. I just figured that was the way of the world. A sort of "divine detachment" as the phrase goes. An acceptance that some level of pain and suffering is a necessary part of existence — as Nietzsche said "Have you said Yes to one joy? O my friends, then you have said yes to all woe as well! All things are enchained, all things are entwined, all things are in love!" .... Once you take that point of view to heart fully enough, maybe you stop worrying about the particulars — whether one or two (or one or two million) extra little girls get raped and murdered is kind of a rounding error...

Also there's the possibility that we're part of a giant data mining experiment, right? The intelligences who created us could have created a vast number of universes containing a vast number of intelligent species and civilizations, and then they could be studying how each of them evolves. Some may have more pain and suffering, some may have less —- they may have introduced a lot of variation to make it an interesting study!

There are many possibilities to think about, and many that seem (to me) more naturally implied by the evidence than the notion of a creator who genuinely wants to minimize our suffering...

Lincoln

You mention that you don't see why humans would have to be less benevolent than any posthumans that created us. Nothing starts as benevolent. It's learned, even if that learning takes the form of executing copied code. The knowledge to write the code came from a process, and the process of executing the code reflects the originally learned process. How does that process feel? Does it suffer? Could it do otherwise (as your Nietzsche quote points out)?

So why couldn't or wouldn't posthumans make the process faster and more direct? Why would we have to evolve through environments that would initially select for violence? If they intended us to become posthumans, why not just pop us out fully formed? Well, it seems to me that either they don't exist (which I reject for practical and moral reasons, as well as esthetic reasons), or there is some additional benefit to the slower and indirect process of becoming posthumans (maybe data mining, as you point out), or we're overlooking some other assumption. What?

This returns us to the problem of evil, as well as to other comments you made in your feedback. You mentioned you think it unlikely that any creator of our universe prioritizes minimizing our suffering. I agree, and that squares with Mormonism, which posits God's work and glory to be bringing about immortality and eternal life (this latter understood as the quality of life God lives), rather than something like minimizing suffering. If God's priority were minimizing suffering, his work and glory would be to make everything dumb as rocks.

Let's say God exists and, in addition to whatever data mining benefits God may get (which, incidentally, works well with the idea in Mormonism that God continues to progress), we're also benefitting from the slow indirect process of learning benevolence through suffering. What would be the benefit? Could we get it some other way? Maybe there's value for each of us in the diversity of benevolence that arises from the diversity of suffering – indefinitely numerous unique aspects of the divine virtue? If so, that would feed into one of the more common Mormon interpretations of the Atonement of Christ, which holds that Christ suffered everything that each of us has suffered. If benevolence arises from violence, what could be more benevolent than a God that chooses to suffer in all the ways that all of its creations suffer?

In Mormonism, there's also the paradoxical idea that, on the one hand, we need not suffer so much if we will simply choose to repent and allow the Atonement of Christ to make us clean from sin; on the other hand, if we would be joint heirs with Christ in the glory of God, we must take on the name of Christ and participate as saviors in the suffering. As I interpret this, there is a communal complexity at play, wherein we each participate differently according to our circumstances, desires and abilities. The end result is not a simple uniform deification where every person is alike, but rather a complex diverse deification where, to paraphrase Mormon scripture, we enjoy the same sociality that we now enjoy, except that it becomes coupled with divine glory.

I suppose the idea is that there are no shortcuts to becoming like God – to becoming part of God – to becoming God. There is no other way, brutal though it may be at times, to attaining the rich complexity of shared divinity. There is no compression algorithm. It must be lived. It must be chosen. It is only that which we all bring to it.

I expect we'll start to see these principles come into play with increasing force in the work to develop artificial intelligence. If an AI never learns how to do long division, it won't know what it feels like to do long division. It won't empathize as directly with those who know the feeling. It won't laugh or cry as authentically when someone reminisces of those good old days when division required an emotional effort. It's kind of trivial, on the one hand, yet kind of amazing on the other, this cosmically nearly-insignificant period of time during which humans on Earth performed long division. Now, I'm not saying an AI couldn't learn long division. Rather, I'm saying a divine AI, a benevolent AI, insofar as a few humans are concerned, would have to learn, and learning would have to feel the same, and feeling the same would be experienced the same . . . and this is where we get into the challenges of identity that advanced AI will present. Maybe we're already playing them out now. Maybe we are part of an advanced intelligence, learning what it's like to be human.

Ben

So part of your idea is that if we want to become God(s), we need to experience everything that God(s) have experienced, including a lot of suffering? A mind that has not known sufficient suffering can never be a truly Godly mind?

Lincoln

That's clearly part of the Christian narrative, emphasized in the Mormon tradition as "God himself" suffering for and with the world. I'm not saying that we should promote suffering or engage in anything like self-mutilation. I'm saying, though, that empathy, which is at the heart of benevolence, does seem to arise in us, both individually and as a species, from our own difficult experience. I also doubt that it's possible to develop an artificial intelligence that humans would recognize as their equal or superior without incorporating that which would logically necessitate the capacity to suffer. What justifies our development of AI in a world of tension and conflict? What justifies bringing children into the world, knowing they'll suffer? Our reasons vary, but almost none of us do it for the suffering. We do it because the overall value proposition is worth the risks, whatever the imagined opportunities may be in all their diversity. In Mormonism, there's an idea that we and all living things experience joy in filling the measure of our creation. I don't think this is an appeal to any narrow sort of happiness, but rather something broad to the point of including even the kind of joy experienced in moments of real courage.

Ben

Oh, and one more thing I just can't resist asking... In your perspective could an intelligent computer program have a soul? Could it have consciousness? Could an intelligent computer program become a God, in the same sense that a person could? Will AIs be able to participate in the collective deification process on the same status as humans?

Lincoln

Technically, in Mormonism, "soul" is used to describe the combination of spirit and body, rather than just spirit. That aside, my answer to your question is that I think computer programs already have spirits, or actually ARE spirits.

Ben

Hah – perfect!

Lincoln

In Mormon cosmology, God creates everything spiritually before physically, organizing and reorganizing uncreated spirit and matter toward greater joy and glory. All things have spirits. Humans have spirits. Non-human animals and even the Earth have spirits, and will be glorified according to the measure of their creation, along with us. Many Mormons also anticipate that the day will come when, emulating God, we learn to create our own spirit children. Spirit is in and through all things. Recall, too, that Mormons are philosophical materialists (not dualists), so even spirit is matter, which God organizes as the spiritual creation of all things. So far as I'm concerned, spirit as described by Mormonism is information, and software engineering is spiritual creation. We are already engaged in the early stages of the creation of our spirit children. Taking a step back, consider how this adds perspective to the problem of evil: what justifies our development of artificial intelligence in an evil world?

Ben

Got it. That actually accords rather well with my own view.

Well, thanks for answering all my questions so thoroughly and with so much tolerance. I have to say I'm not particularly tempted to convert to Mormonism even after hearing your wonderfully subtle views – but I do have a much better appreciation for why you think Mormonism and transhumanism are compatible. And I am impressed with the flexibility and robustness of the Mormon belief system, that lets it adapt and reinterpret itself to encompass transhumanist technologies, without losing sight of its essence at all. Very, very interesting...

Lincoln

Ben, thanks for making the time to discuss these questions with me. While I'd certainly welcome you or anyone else as a convert to Mormonism, more important to me is the work to improve mutual understanding and to make a better world. In fact, I hope none of my words give anyone the impression that I think everyone must be Mormon to make that better world. To the contrary, I see something beautiful, even divine in my estimation, unfolding all around and within us, in part as a direct consequence of our diversity of religious and non-religious perspectives. Early Mormon leader Orson F Whitney put it this way: 'God is using more than one people for the accomplishment of his great and marvelous work. The Latter-day Saints cannot do it all. It is too vast, too arduous, for any one people.' I believe that, and in part such ideas are what motivate the Mormon Transhumanist Association.

Natasha Vita-More: Upgrading Humanity

Interview first published January 2013

I first encountered Natasha Vita–More's thinking in the 1990s via her work as President of the Extropy Institute – the first serious transhumanist organization, which has played a large role in setting the tone for transhumanism and modern futurist thinking. Extropy made a strategic decision to close its doors in 2006, but Natasha has not slowed down her work toward extropic goals at all; in fact she has expanded the scope of her transhumanist work in a variety of directions. I've had the pleasure of working with her for several years now in the context of the transhumanist advocacy organization Humanity+, of which she's currently Chair and I'm Vice Chair.

Much of Natasha's work is focused on human enhancement and radical life extension – or what she has sometimes called "life expansion." She brings a combination of artistic and design sensibility with conceptual and academic sophistication to her endeavors, such as her pioneering future human design "Primo Posthuman." Primo Posthuman proposes what she calls a "platform diverse body" and "substrate autonomous identity." The ideas underlying Primo Posthuman are plausible and fascinating, and are drawn from the scientific and speculative literature. What Primo Posthuman accomplishes is to wrap these ideas up in a visually appealing and conceptually simple package, getting the idea of an advanced enhanced human across viscerally to seasoned transhumanists and newbies alike. The design is ever-evolving and Natasha is now preparing its 2013 version.

Natasha's written and graphical work has appeared in numerous magazines; her multi-media creations have been widely exhibited around the world; and she has appeared in over twenty-four televised documentaries on the future and culture. In addition to her role as Chair of Humanity+, she is fellow at IEET and the Hybrid Reality Institute, a visiting scholar at 21st Century Medicine, and a

track advisor at Singularity University. She holds a PhD from the Planetary Collegium at the University of Plymouth, an MSc in human enhancement technologies and an MPhil on the theoretical concerns of human enhancement, and is a certified personal trainer and sports nutritionist. She is currently a Professor at the University of Advancing Technology.

It was my pleasure to interview Natasha on her work and thinking, with a focus on her Primo Posthuman design and its successors, and the concepts underlying these designs.

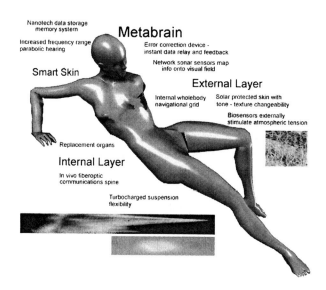

Ben
You've stated that "in most instances the cyborg lacks social consciousness of the transhuman and suggests a grim and dire nature by impersonalizing humanity[193]."

What do you think are the reasons for this? Why did the early notion of a cyborg develop in this way?

Natasha
Manfred Clynes and Nathan Kline developed the concept of the cyborg (1960), it was intended as an augmented human body that could perform in zero G for space exploration. It was not a particular entity with agency, as later suggested by Kevin Warwick or Steve Mann. It was not a metaphor for a feminist agenda, as later suggested by Donna Haraway. While Warwick, Mann and Haraway all added interesting physical and metaphorical interpretations to the concept of cyborg, in actuality the cyborg was and continues to be an appendage to human body. The cyborg originated out of cybernetics and space exploration and was not considered primarily a means for the human species to go off planet to generate new types of intelligences. The idea of space exploration had a more practical mission: mining asteroids, competing in the Western world's quest for space dominance, etc. From this understanding, the cyborg is something we do to the human body as an end result-- making it suitable for space living in a zero G environment. This idea contrasts the self-directed evolution of the human in becoming something other, such as a transhuman, posthuman or upload (SIMS).

Ben
Hmmm... and what about the future? As we progress further toward actually realizing advanced cyborg technology, do you think the general vision of cyborgs in the media and mass

---

[193] http://www.natasha.cc/CiberART%20-%20Primo%20Posthuman%20Future%20Body%20Design.htm

consciousness is becoming MORE socially and transhumanistically savvy? Or less so? Or, no real change?

Natasha

Until recently, the term "cyborg" has had more historical currency in the mainstream than the term "transhuman". And while the cyborg is more physically frightening to look at than a transhuman, it has become a familiar concept and people often accept the familiar, no matter how daunting it is, in preference to the unknown. After all, the cyborg is associated with Hollywood filmmakers, actors and highly acclaimed science fiction authors. But this is changing. Over the past five years the notion of the transhuman has more pragmatic currency due, in large part, to the established domain of life extension. The popularity of life extension has affected the awareness of alternative human futures because, specifically, one of the underlying aims of the transhumanist community is life extension. And transhumanism also has contact with two other subfields that have huge attention in the public and private sectors: evolutionary biology and cognitive science, which gain attention through diverse debates on ethics and human futures.

Ben

There is a very rich and deep interplay between the human body and the human mind and self, as you know. I very much enjoyed your conceptual and graphical "Primo Posthuman" design for a functionally and aesthetically upgraded cyborg body. Let's suppose, hypothetically, that we were able to port a human mind from a legacy human body into a *body with a diverse physical substrate, such as your original Primo Posthuman, or any newer version of the concept*. How do you think this would affect the self-system of that mind, and its ways of thinking? Do you think it could effect a radical personality transformation, so much that it would hardly be the same "psychological person" after as before? Or would the changes be more subtle? Or do you reckon it would depend strongly on the individual?

Natasha

Exploring new environments affects our sense of self and either adds to our level of confidence or breaks it down, the latter due largely to unintended circumstances and stresses that cause reticence and confusion. Nevertheless, we continue to explore and challenge our sense of self. If we port a human mind into a platform diverse body like Primo Posthuman, we have to also transport other aspects of the brain, which include the central nervous system and an array of sensorial experiences. These memories provide some sense of familiarity to our personhood. Without these aspects, we could become mentally discombobulated and split off into varied personas that could experience types of uneasiness or in some cases, psychosis. All in all, continuity of identity, or a diachronic self, is crucial for a future human mind encapsulated in a new body. My view is that we can change who we are by changing our focus, environment, skill set, attitude, etc. and transforming our personality with a proactive strategy. It is not easy and it takes work, but it is doable.

But let's take the scenario of a mind transfer into a nonbiological platform: In order to do this, the engineering of identity transfer needs to be far more advanced. Not only this, it must be more transdisciplinary– it is not just an advance in technology but also an advance in our consciousness – our awareness of this field and how it affects us. So, in short, the bandwidth of information about mind transfer/identity transfer would have to build a broader arena for discussions for understanding and encourage corresponding fields such as experience design to be on board. For example, experience design would become a major player in providing an "experience" of what it might be like to be in a different body. Experience design would team up with multi-media technologies such as gaming, immersivity, smart architecture, etc. Through new experience we can changes what we think about our future and go through processes (preparation, coaching and tutoring) of what it might be like to be an upload. Gaming and HCI interaction could be used as a means to help us to take baby steps in getting an "experience" of what it might be like. This example takes the field of mind transfer and identity into a broader spectrum of approaches.

Ben

Yeah, I think I get it.... We could create video games, virtual reality experiences and so forth, which were specifically intended to help people get used to their new bodies, before they start actually wearing the new bodies. This would make it much less of a shock and encourage "continuity of self", which is critical.

It would also be great, I suppose, if you could try on the new body for a brief period, and then go back to the old one. In that way there could be a gradual transition psychologically, even to a radically different sort of body...

One other thing that strikes me on reading about *your concept of platform diverse bodies and whole body prosthetics (such as Primo Posthuman)* is that, with all the advanced technologies it assumes, there's really no reason for it to assume such a humanlike form, right? I suppose the assumption is that many humans will prefer to retain a humanlike physical form even when that's not strictly necessary in terms of the technology underlying their (engineered) bodies. But I wonder the extent to which this is actually true. I don't think I'd personally have much desire to retain a humanoid form, if it weren't really necessary – I guess I'd end up some kind of shape-shifter if possible. So I wonder if you've thought about nonhumanoid designs conceptually similar to Primo Posthuman – maybe a bird-man, a dolphin-man or some sort of space amoeba?

Natasha

I intentionally selected a human life form rather than mechanized structure. My strategy was that since most of the future human concepts were overtly metal and machine, that there was a gap in the visual aesthetics of what a future human might look like. Since "metal" and "machine" suggest a cold, hard, non-feeling aesthetic, I wanted to introduce a humane characteristic that symbolizes our human strength, perceptibility, intelligence, and sensuality. Within these strata, I incorporated the element of mutability. By this I mean that even though the original concept of Primo Posthuman looks human-like, it encompasses a real-time aesthetic that provides a "Metabrain" appendage for uploading. In other words, the physical form is a real-time functional machine that incorporates the humanness we humans are fond of and offers a means to travel in non-real time settings.

I am not excited about having fish fins or bird wings, although I have enjoyed many of these styles in Second Life. My aim is to design a body that is sustainable and streamlined. After that, it certainly could have all sorts of appendages (like Laura Beloff's "Tail") or a simulated exoskeleton for exploring diverse environments. But for the purpose of a transhuman evolution and the fact that we exist now in real time as biological beings, we need to be practical first, and then see where that takes us. In SL or other simulations, we can use an array of designs (like Elif Ayiter's avatars).

Ben
OK, so about being practical... Maybe this is a stupid question but: If given the chance to upgrade your physical body step by step into a Primo Posthuman type condition, would you do it? What do you think the process would feel like? How would you react to it psychologically, do you reckon?

Natasha
I have been upgrading my physical body by following health protocols for longevity that directly affect my body, brain and mind. Specific upgrades include artificial lens on both right and left eyes, hormone replacement therapy, and varied interventions for dealing with cancers, etc. As far as a posthuman body concept, I don't have any of the specifications that I write about, but it's not an option today to exchange a seemingly healthy bio body part for an upgrade part.

Until then, I have a series of bio-basics for upgrading. I am a certified personal trainer and sports nutritionist, spent years at the Yoga College of India, and am trained in TM (Maharishi Mahesh Yogi). Combining this knowledge, I focus on aerobic and anaerobic exercise routines, Yoga, and meditation. Brain plasticity is imperative, so I work at learning new skills as way of life, both intellectual and physical. I also think that sensuality has a lot to do with attitude, appearance and bodily wellness -- so sex gets a high rating. Laughter is equally as consequential, so I have my favorite comedians I watch and TV shows that are mental ticklers. But let's take a look at a scenario where I would need a new limb or cognitive add-on.

To replace a worn out bio limb with an upgraded AI-driven robotic smart part would be extremely exciting – like driving a Maserati luxury sports car with unmatched style and performance instead of a Ford Escort with roll down windows. Big difference. The Primo Posthuman body would drive smooth, flexible, durable, with extreme usability. So, what would it feel like? A little tricky at first and I'd have to acclimate. After a trial run, it would be amazing.

Ben

Your work and your thinking on these topics is, obviously, extremely interdisciplinary. That's one of the things I enjoy about your work. Could you reflect a little on the importance of keeping people with a design and arts mindset involved in creating advanced technologies, such as cyborgs? And on the value of having cross-disciplinary education and culture, so that there's not a rigid divide between the scientists/engineers on the one hand, and the designers/artists on the other. How critical do you think this is for the creation of an exciting, desirable, productive future for humanity and our transhumance descendants?

Natasha

The old art vs. science issue was handled very nicely by C.P. Snow (i.e., *The Two Cultures*) back in the mid-1960s. But I give a lot of credence to Susan Sontag's assessment of how certain constructs try to make intelligence more meaningful than aesthetics. She didn't pontificate about cultures, and in her book *Against Interpretation and Other Essays* she tackles the issue quite nicely by suggesting that our interpretation of the world around us is often coated with biases of universal truths that, in essence, are not truths at all. She pointedly argued for erotics of art over hermeneutics. But the overall scope of creating is about innovation and those who innovate are solving a need and filling a gap, which is not sequestered to one discipline. John Brockman was wise in crafting the concept of a "third culture" because he recognized the enormous crossover between science and creative thinking.

Nevertheless, why people keep trying to make an issue out of art versus science is interesting. I suppose it offers a distinction between what we think is an intellectual and practical activity to try to understand the world around us, and what we think is not. The aim of art, as I see it, is to understand the world around us and reinterpret it through varied approaches – gaming, filmmaking, fiction, painting, sculpture, bioart, virtuality, music, etc. The central difference is that science tries to do this objectively and art tries to do it subjectively. However, there is a tremendous gray area between what is objective and what is subjective and they often cross over. A scientist could say her work is an objective search for knowledge through but is deeply influenced by her ideology or religious beliefs. Likewise, an artist could say his work is subjective, but he has an objective aim in mind – a goal and is driven by this. Design is a different matter. Design is all about problem solving and doing it with an extreme level of elegance and excellence, much like mathematics – it is closely related to art, but not the same.

So, if we look at the divide as a non-issue and realize that the last decade has been the spawning ground for a stronger, broader confluence of fields and that the old world academic narrative of separating fields and disciplines has evolved into a type of hypermodern world of cross-disciplinarity, multi-disciplinarity, trans-disciplinarity, openness, and collaboration –

then I think we can handle any schism with grace and a warm handshake.

Ben

In terms of practical R&D work going on today, what excites you the most in the context of *progress toward platform diverse bodies and substrate autonomous persons*? What gives you the greatest feeling of real practical progress toward the goal of realizing Primo Posthuman or something like it?

Natasha

The fact that the maximum human lifespan is limited to a little more than a single century, most of which is spent resisting disease, is the main thing that compelled me to conceptualize a future body prototype. As a teenager I volunteered at The Home for Incurables, a facility where people were so malformed that they were not allowed to go out in public. At St. Jude's Children's Hospital, I volunteered and observed children with incurable cancers awaiting their ultimate and untimely death. These and other experiences over the years are a vital part of my awareness that I cannot ignore. I have an unfaltering interest in future possibilities of improving, protecting and sustaining life. Because of advances in methods to intervene with the damage caused by cellular mutations and aging and to repair the body, there is reason to be more enthused than ever to explore how these emerging and speculative technologies fit into creative explorations in the arts and sciences. Bio-design is making headway and its cousin, bio-art, has become a substantial field with the arts. DIY-bio is more involved with HCI, but it does bring in a wide spectrum of ideas for creative explorations (not always safe or smart, but interesting nonetheless). I prefer the combined efforts of bio-design and bio-art because they cover two aspects of creative innovations: artistic pursuits that are uniquely imaginative and design-based pursuits that are focused on problem-solving.

The most basic practical purpose of a future body prototype like Primo Posthuman or my other platform diverse bodies is to give people a body that is more sustainable than their pre-programmed time-limited bio body. This aim matches my initiative to design and build a body for humans who suffer from physical disease and need a new or alternative healthy body and, specifically, to resolve the issue of repair and revival of cryonics patients. Ultimately, many neurosuspension candidates at Alcor and other cryonics facilities will need a new body if and when we are revived. My prototype is designed with this in mind.

Ben

How have the Primo Posthuman design and idea, and its successors, been received? Has there been a difference in the sort of reception this line of research has gotten from scientists versus arts people? Have you been deluged with people asking where they can buy one?

Natasha

Amazingly, I'll answer this with a question: How many times in our lives do we have an idea and build on it and then someone sees it or reads about it and it goes off like fireworks around the planet? Well, that is what happened with "Primo Posthuman". It has been on the cover of dozens of international magazines, featured in numerous newspapers and televised documentaries and continues to spark media attention. It is surprising because for years I had to deal with negative press on issues related to radical life extension – due to peoples' fears about overpopulation, a psychology of selfishness, a lack of care about the environment, or only the rich obtaining the technology, etc. Contrary to people's lack of knowledge, the field of life extension has a profusion of research that cover all sides of the issues. Fortunately, "Primo Posthuman" (and its multiple versions and revisions) has directly influenced how people think about human enhancement and human futures. The on-going enthusiastic reception of my work has been even more rewarding because I have been able to make a dent in the rigidity of fearful thinking and have offered a possibility – a glimpse of what could happen to the human body and mind. And since prosthetics are more and more seamlessly interfacing with robotics and AI, and prosthetics has resurfaced as a stylized addition to the body rather than just a way of replacing a damaged part, it looks like it could take off and engage concepts such my idea of a whole body prosthetic.

Ben

What are you working on these days? Anything that dramatically enlarges or drastically revises the basic Primo Posthuman vision you laid out previously? How has the Primo Posthuman related concept developed since its early days?

Natasha

The most recent rendition of my original work has surfaced as a type of platform diverse body and a substrate autonomous person. I focus on personhood rather than mind because of issues of continuity of self. The platform diverse body is a confluence of organic and computational systems and is a primary means for life expansion. If medical science can reverse aging and help to keep the bio body functioning indefinitely, that is great but it does not resolve the problem of unexpected death. It might be wise to continually backup our personhood (brain functions and especially memory and identity) and also have a second body just in case.

Since my design includes mind transfer (uploading) as an option within the Metabrain, the design serves as a both transport vehicle in real time and a transport devise for persons (consciousness) to exist in simulated environments. I am working on new designs that expressly emphasize this duo-purpose.

I'm also working on some other exciting ~~stuff~~, media-related works not so directly related to Primo Posthuman. For instance, one amazing project I'm working on is a scientific research project of c.elegans in association with mind and memory. This is slated to take place at a leading research facility. The outcome would be available for museum exhibition, including filmed documentation. But this is under wraps for the moment and it's probably best not to talk about it before it happens.

A favored project is my H+ TV series which draws from my cable TV show in LA and Telluride in the 1980s/1990s and which seeks to fill a gap in knowledge about science, technology and design and radical life extension. You can see the tease here: http://vimeo.com/49258226. I hope to continue to work with teleXLR8 as a venue for it, but also with several universities to bring in students to participate in the discussions and debates.

I'm excited about a book coming out this year *The Transhumanist Reader: Classical and Contemporary Essays on the Science, Technology and Philosophy of the Future Human* (Wiley). I co-edited with Max More and we both are contributing authors of its collection of 40 high-profile authors and seminal ideas that helped to form the scope of transhumanism. I'll be spending time promoting the book.

Ben

All fantastic stuff indeed! ...

Now... this question is a bit different from the others, but I'm wondering, what do you see as the relation between postmodern philosophy and transhumanism? I feel like I see a few shades of postmodern thinking in your work – and of course, contemporary design and arts have been greatly influenced by postmodernism. And yet, postmodernism in art and philosophy often tends to go along with a certain directionlessness, almost a sort of nihilism – whereas your view of the transhuman is largely about extropy and purposeful, rational, goal-oriented growth.

Natasha

While I value Ihab Hassan and admire his contributions to knowledge from the postmodernist perspective, in general, I think postmodernism is a cliché. Its narrative fights against universals, scientism, etc. and offers little resolve. It is often more of a reaction than an action. The benefits off the postmodern agenda that I recognize are its deconstruction of universals and ardent efforts to encourage a feminist agenda and gay and lesbian awareness, and transcend the ivory tower of scientism, for example. But it offers nothing really tangible outside of smashing other philosophical worldviews. It wants us to disengage from our past – to toss everything out without discerning what has been beneficial to human knowledge and what has been harmful. For example, humanism has some bad points, to be sure, but it also has come important concepts. Why be antihumanist? That seems a bit reactionary and not taking into consideration the overall period within history. I do not support humanism, mainly because it is more focused on "man" than "human" and it lacks a sophisticated understanding of human evolution (although that is changing I hear). But why bother to attack it? Overall, I think postmodernism is a pain in the rear-end, not to mention the postmodernist writing style. I hope there are no traces of postmodernism in my work, and I certainly have not encouraged any. Rather, I have sought to be free of "isms" I my work and create from my own life experience.

Postmodernists have tried to create the posthuman and posthumanism as their sword against the future and I think have failed. There is a lack of visionary thinking. but let me say that the postmodernist scholars of posthumanism are highly influential and highly accomplished in their own works, when they attempt to sort out the posthuman they become confused and biased. I think the book H+/H-: Transhumanism and its Critics evidence this in large part, although the final comments were allotted to the postmodernists and not the transhumanists. But at least we had an opportunity to intellectually wrestle with them.

Where transhumanism and posthumanism part ways is in their respective approaches: Posthumanism is deeply intertwined with the postmodernist agenda and rhetoric while transhumanism is deeply intertwined in human evolution, human enhancement and life extension. This difference brings into the discussion how or in what ways have posthumanist thinkers offered concrete concepts for which to deal with the human future? Where is there a posthumanist view that has developed a strategy for dealing with the issues that surround human futures? Where is the posthumanist struggle that seeks knowledge and practice outside theorizing that can make a difference? Alternatively, what frames of reference might best form methods by which to project possible futures?

The relation I see between postmodern philosophy and transhumanism is the desire to move beyond universals and to encourage more diversity amongst humans. We need to ask ourselves if we accept the dystopic postmodern view of progress or if we move forward in our evolution in becoming diversified agency with new types of bodies with a sense of responsibility and efficacy? I select the latter. We need to liberate ourselves from some of the inevitable limitations people and society accepts within the postmodernist stance. Transhumanism reaches far beyond the postmodernist agenda and has left it in the dust.

I'm not sure how contemporary design has been influenced by postmodernism, outside of architecture. Postmodernism did embellish ornamentation, symbolism and a type of visual wit, but ludics has been around far longer than postmodernism. Benefits of postmodernism in architecture can be seen in a more organic design, rather than modernists' geometric flavor. In fashion, postmodernism allowed for gender-mixed style. In sculpture and product design, an allowance of humor, and the ridiculous is valuable but even here form does follow some type of function. Every art genre borrows from other genres' and makes its mark by critiquing and disavowing its predecessor. That is what makes a new art period, movement or genre novel.

But regarding the wider scope of the postmodernist agenda outside design, in the humanities and philosophy it has ignited a strong fear and negativity within the arts and humanities and persists today.

Ben

Very interesting and passionate response, thanks!

I myself take a somewhat less dim view of postmodernism than many transhumanists – I've learned a lot about time from Deleuze; and a lot about the relation between mind, society, economy and reality from Baudrillard; and a lot about the subtle psycho-social underpinnings of history from Foucault, etc. I suppose the commonality I felt between your work and postmodernism is well captured by your statement that "The relation between postmodern philosophy and transhumanism is the desire to move beyond universals…"

Natasha

I was looking more at the general nature of postmodernism and not specifically at a few uniquely creative minds. Deleuze, Baudrillard, Lacan and Foucault offer substance to philosophy and the arts. But their writings became, in large part, an institutionalized version of themselves for the benefit of an academic postmodernist scope. So, it is not necessarily that their works that were problematic, but often the interpretation and overuse of their works became concretized as "truth" and an end-point rather than their works being an inspiration for furthering knowledge-gathering. Deleuze offers insights into "difference" and the "other". This is important for the transhumanist scope of identity and the mere fact that the transhuman has become the "other", but Deleuze does not reach far enough to concern us with the continuity of identity. He says static with differences, even if in a series of differences. It seems that difference as "diversity". But nevertheless, he spent considerable time arguing against modernity to a point that could have limited his looking forward to transmodernity or hypermodernity. Likewise Baudrillard, for all his insights, negates history by suggesting we have lost contact with the "real". But what about the new real? Anyway, there have been marvelous philosophical minds over the centuries. I prefer Nietzsche and Spinoza.

Ben

Yeah... Whenever discussing philosophy, I always end up coming back to Nietzsche, it seems. The postmodernists pointed out very well that there is no absolute reality and no absolute values... But Nietzsche, commonly cited as precursor to postmodernism, recognized the lack of absolute values, and then posited that a primary function of the superior mind is to create its own values. In this sense we could say that transhumanism is creating its own values, centered around notions of progress and expansion... Postmodernism certainly allows for the possibility of creating subjective meaning in this way, but chooses not to focus on this... And in the end, what one chooses to focus on is perhaps more important than what one acknowledges as possible. I love where transhumanism focuses its attention – on creating better and better futures – but I think postmodernism also provides a service via focusing attention on the groundlessness of it all, on the psycho-social construction of assumptive realities.

Natasha

Well, transhumanism certainly has a great sense of ludics and irony and is also grounded in its philosophical tenets. The great parties, artistic projects, science fiction stories, critical assessment of knowledge, heady debates, etc. are all endemic to transhumanism. I cannot see anything baseless about it, and certainly not about the current lack of plurality of death. But there are different flavors of transhumanism, so can see why someone might find a bit of a bad flavor among some people who self-define themselves as transhumanist, as well as amongst those who self-define themselves as postmodernist. Not everyone is ethical or values life equally.

And I suppose this is one of the great benefits of the arts – architecture, fiction, design, filmmaking, gaming, etc. – that these modes of reality offer a multiplicity of experience ("no absolute values", as you say) So, it could very well be that an experiential cognition rather than just a reflexive cognition allows for the varied, unbounded existences for all to explore.

Ben
Well, we could certainly continue to discuss these points for a long time, but I think you answered my question very well. Philosophical discussions do tend to keep going around and around and around, whereas the nice thing about transhumanism and allied technologies is that they keep moving forward – at least they turn the circling-around of philosophy into an amazing variety of helices and so forth!

Like you, I'm definitely looking forward to an expanding and increasing multiplicity of experience in future – hopefully a very long future for both of us!

Jeffery Martin & Mikey Siegel: Engineering Enlightenment

Interview first published August 2012

In 2011 I shifted to spending most of my time in Hong Kong, and since that point I've met an amazing collection of fascinating people here – some HongKongese, some mainland Chinese, and some Westerners or other Asians passing through. Hong Kong is the sort of place where all sorts of folks will pass through for hours, days, weeks, months, or years – and occasionally stay a lifetime. Among the more stimulating encounters I've had here, was my intersection with Jeffery Martin and Mikey Siegel, and their work (in collaboration with Gino Yu) on

- the psychological and physiological nature of "enlightenment" (or as they often call it, "extraordinary states of well-being" or "persistent non-symbolic states")
- methods for working toward enlightened states of consciousness using neurofeedback and other "brain hacking" technologies.

I learned a lot from talking to Jeffery and Mikey about their research, and was very impressed by their rigorous, clear-minded scientific approach to these confusing and multidimensionally thorny topics.

I was entranced by Jeffery's tales of his travels around the world interviewing rafts of different enlightened people, interviewing them and trying to understand their commonalities and differences. And I was intrigued by Mikey's hacking of EEG game controllers, with a goal of turning them into neurofeedback-based enlightenment engines. We have increasing knowledge of which parts of the brain are differentially active in enlightened minds – so that (or so the idea goes) using EEG equipment to visualize one's brain activity in real-time, one can nudge one's brain-state gradually toward enlightenment, using the visualization as a rough measure of one's incremental progress. There were also some fascinating experiments with noninvasive electrical stimulation of the brain, via carefully placed electrodes, which in some subjects resulted in extraordinary shifts of subjective experience.

We had many deep discussions about the various possible routes to extraordinary, enlightened states of mind. Gino Yu, our mutual collaborator and the one who introduced me to Jeffery and Mikey, was pursuing research on "enlightenment via digital media" – creating video games and multimedia experiences designed to open peoples' minds. I tended toward a frustration with the intrinsic limitations of the human brain, and a propensity to think about mind uploads, AGIs or brain implants. Jeffery and Mikey staked out an intermediate ground of sorts – focusing attention on using advanced technology like neurofeedback and brain stimulation to nudge human brains into better states than their usual ones. Of course there is no contradiction between these approaches; there are many possible paths to many possible kinds of mind expansion.

Jeffery and Mikey have since moved on from Hong Kong, and since their research is still underway, they haven't taken the time to publish their ideas and results yet – though in late 2011 Jeffery gave some great talks on these themes, e.g.

- at the Asia Consciousness Festival[194]

- at TedX Hong Kong[195]

- in a panel discussion at Humanity+ Hong Kong[196]

So I thought it would be interesting to do an H+ Magazine interview with the two of them, as a way of getting some of their

---

[194] http://www.youtube.com/watch?v=-Wt9cBJX8Ww
[195] http://www.youtube.com/watch?v=1kOyrkr4Gtw
[196] http://www.youtube.com/watch?v=Sv5GSexET4o

thinking further spread into the global memeplex, even while their research is still in a fairly early formative stage.

Both Jeffery and Mikey have the cross-disciplinary backgrounds you'd expect from folks involved in this sort of research. Jeffery is a software technologist, serial entrepreneur, scholar, author and educator, with a psychology PhD and interests and achievements spanning many fields including: leadership and management, politics, psychology, computer science, spirituality and religion, electronic media, and personal transformation. His co-authored novel "The Fourth Awakening" explores enlightenment-based themes in a poetic yet rigorous way. When I met him in Hong Kong he was lecturing in the School of Design at Hong Kong Poly U, as well as pursuing various technology/media business projects, writing, and, of course, studying enlightenment.

Mikey graduated from UC Santa Cruz with a BS in Computer Engineering, and then did his Masters at the MIT Media Lab with social robotics legend Cynthia Breazeal. A first-class robotics and general hardware tinkerer as well as a deep thinker, his current interests center around technology and consciousness – as he says, "I'm working to create tools that facilitate people's own path toward higher consciousness, self-realization, awakening, however it is you may describe the experience of transcendence."

Ben

Jeffery, I know you've been spending a lot of time in recent years studying "enlightened" people — or as you've sometimes referred to it, "people who are continually in a state of extraordinary well-being." As I understand it you've traveled the world visiting a host of different people apparently possessing persistent "enlightened" states of mind. I'd be thrilled for you to share some of your findings about these people with our readers. What common factors distinguish these individuals? What makes them different from the rest of us?

Jeffery

That's a great question Ben, and of course it goes to the heart of all of our research. The reality of course is that there are many things that differentiate them from the rest of the population. This is one of those types of questions that can be answered on many different levels, from the level of personality to the level of brain function. Perhaps the most relevant level is how it changes one's experience of the world. It seems to bring with it a deep sense of inner peace and also a sense of overall completeness. Now, it's important to note that this really depends upon where someone is that in regards to the overall process. One of the most important things that we've learned is that this notion of enlightenment or "extraordinary well-being" is not a binary state. It's not something that is simply turned on and then is the same for everyone. There's actually is a range or continuum — and where someone falls along it determines the experience that they have in relation to the world.

Ben

Yeah, that makes total sense. But still, there must be some commonalities, regardless of where someone falls on the continuum. Continuum or not, all these folks you're studying are qualitatively different from, say, the average teacher I had in high school.

Jeffery

What all of these people have in common, regardless of where they fall on the continuum, is the experience of a fundamental and profound shift in what it feels like to be them. In other words — a significant change in their sense of self. Overall, this seems to involve a shift away from a tightly focused or highly individuated sense of self to something else. What that something else is, often relates to ideology or beliefs the individuals had before this happened to them. So for example if someone is a Buddhist it may show up as spaciousness. If someone is a Christian it may show up as a feeling of God, Christ, or the Holy Spirit indwelling them. As individuals progress downs continuum often even this falls away, but the reality is very few make it that far. If they do, however, they generally find themselves in a place that is beyond the visceral experience of their previous ideologies and belief systems.

Ben

I wonder about the extent to which the actual experience of enlightenment differs depending on the person's belief system before their transformation, versus the difference just lying in the way they describe their enlightenment? A Christian may describe their experience in terms of the Holy Spirit, whereas a Buddhist may describe it in terms of the pearly nothingness – but actually their experiences might be quite similar, apart from the way they label it and connect it into their prior belief systems...

The psychological nature of the experience may be more invariant than the language used to describe it – do you think?

Jeffery

In some respects, yes. There are huge, fairly universal impacts on things like emotions. When someone first crosses over, if they enter the low or beginning part of the continuum, there's an immediate emotional benefit. Events that occur in their life still produce emotion in them but these emotions exert much less influence over their moment to moment the experience. One of the ways this manifests is via emotions lasting less time.

As one approaches the far end of the continuum, negative emotions fall away and positive emotions become increasingly highlighted. Eventually these individuals reach a point were they are only experiencing one highly positive emotion.

As they move even further down the continuum and encounter the place where the deeper components of their previous ideologies fall away, emotion falls away entirely. This may sound troubling, but it is replaced with an even greater degree of well-being. And, remember, these were individuals who were already experiencing a level of well-being that is far greater than what most people can imagine.

Ben

Another question that often pops into my mind when I listen to you talk about this stuff is: how would you define a "persistent non-symbolic state"?

Personally I'm tempted to conceive of it in terms of "addiction." I think of addiction, crudely, as wanting something not because of its current appropriateness, but because one had it before. When Buddhists say "non-attachment", I tend to think "non-addiction", broadly speaking. So in this vein, I'm tempted to think of "non-symbolic consciousness" as "the absence of addiction to any particular symbol." That is: obviously enlightened minds can do symbol-manipulation. But maybe they don't get addicted to particular symbols, and particular habits and patterns of symbol manipulation. Instead they use symbols as appropriate, just like they use utensils to eat as appropriate, and then put them down when they're done.

Anyway that's my own — maybe very incomplete, silly or eccentric — way of conceiving "non-symbolic consciousness"... But I'm wondering about yours, since you've obviously thought about it a lot more!

Jeffery

This is a very interesting view, and certainly one way to view it. Personally, I think of symbolic consciousness more as a habit that forms from around age 2-3 onward, and is strongly reinforced by society. Though certainly there must be reinforcement from the reward centers in the brain involved, making your reference to addiction relevant...But I think more in a downstream kind of way.

I define a persistent non-symbolic state as a persistent shift in an individual's sense of self away from a strong sense of individualization, and towards something else. That something else if often ideologically driven. So for a Christian it might be towards a feeling of union with God, or Christ, or the Holy Spirit depending on their sect. For a Buddhist it might be towards a sense of spaciousness or pure consciousness, again, depending on their sect. For an atheist it might be more towards the feeling of greater connectedness to nature.

As this sense of a strongly individuated sense of self becomes less prevalent, other changes accompany it. We've classified these along in the domains of cognition (thinking), affect (emotion), perception and memory. Changes in cognition, for example, initially involve a decrease in the ability to self-referential thought to pull the individual in. On the far end of the continuum, it seems to involve an absence of these thoughts altogether. In between there is a kind of fading out.

Ben

Understood — that makes total sense to me. I'm also reminded somehow of Ray Bradbury's quote on symbolism in writing...

> *"I never consciously place symbolism in my writing. That would be a self-conscious exercise and self-consciousness is defeating to any creative act. Better to get the subconscious to do the work for you, and get out of the way. The best symbolism is always unsuspected and natural. During a lifetime, one saves up information which collects itself around centers in the mind; these automatically become symbols on a subliminal level and need only be summoned in the heat of writing. "*

Similarly, it may be that folks in a persistent non-symbolic state are, in some sense, doing implicit symbolic inference — like Bradbury in his state of unconscious creative flow — without self-consciously creating or manipulating symbols. The relation between enlightened states and creative flow states is another big, deep, tricky topic, of course!

Jeffery

This could absolutely be the case, certainly I've speculated along these lines as well. This is a great quote that I was unaware of, thanks for passing it along!

Ben

Shifting gears a bit – can you say a little bit about what got you interested in this area of research in the first place?

Jeffery

I was researching well-being. We'd come across data in our experiments that suggested that it might be linked to the degree and type of emotional and cognitive release an individual could achieve. Over several years, a pattern emerged that seemed to be a continuum. On 'low' or far left end of the continuum, if you want to think of it as a visual line, individuals seemed to be deeply embedded in their thoughts and emotions. Their sense of self was almost entirely constructed within those internal experiences. As one progressed to the 'higher' or right side of the model, this became less true. After quite a bit of research, we reached the point where we were able to match people up to where they were on the continuum, provide the appropriate tool to move them 'further along,' and so forth. But eventually we hit a point where we weren't making any progress.

At that point we started looking for exemplars that might indicate where the far end of the continuum lied. In the end, only one group of individuals made claims that seemed to fit so we began to research them. This is the group that we've been referring to as enlightened or as having extraordinary well-being. As our research on them progressed it became clear that specific brain regions might be responsible for these persistent changes in how these individuals experience the world. There didn't seem to be any reliable methods that induced it, so we began to explore neurofeedback and also neurostimulation technologies in an effort to see if we could find a reliable way to get people there. That's the effort that we are most engaged in right now.

Ben

What about you, Mikey? Your previous research was on social robotics – the neurofeedback of enlightenment is a bit of a shift. What motivated the change in research direction?

Mikey

My work in social human-robot interaction focused on the ways in which robots can influence human belief and behavior. For me this ranged from maximizing the donation a robot could solicit, to increasing car safety by interactively modifying driving behavior. This propelled me down a broader line of inquiry: what is the most profound way in which technology can influence human experience? The answer I came to was maximizing well-being. That answer led to a more philosophical question: what is the absolute pinnacle of human well-being? This led me down the path of researching the relationship between technology and spiritual enlightenment.

Ben

Much of my own interest in your work comes out of various experiences I've had myself, getting into unusual states of mind via meditation, and various other practices. But I wouldn't consider myself anywhere remotely near "enlightened" – though I've definitely had some pretty extraordinary states of pretty extraordinary well-being here and there, including some I couldn't come close to describing in words.

So, getting back to what I was asking about before — this makes me wonder: What are your thoughts on the general, habitual differences — neurally, cognitive-process-wise, and in terms of subjective state of experience — between A) "enlightened" people, B) highly experienced meditators, C) ordinary schmucks like me?

Jeffery

Well, I wish the "ordinary schmucks" of this world had even a fraction of your brainpower and insight! Obviously, this is a huge topic, because there are large quantities of research into all three of these groups at this point, though more so for the last two groups. Let me narrow the question down a bit. I would say that one of the difficult things in our research is separating out the effects of meditation from the underlying "Persistent Non-Symbolic Consciousness" experience. Meditation produces all kinds of interesting cognitive process effects that are often not related at all to PNS. Just the process of separating these out took quite a while for us, because initially virtually all of the neuroscience research that we had to work with involved meditation. For political reasons within the academy, this is still pretty much true for research that is being publically done. Academic work on PNS is centered around meditators because that's their only hope of getting it published. The private research is typically the opposite, though, so it has been incredibly helpful in determining what are meditation effects and what relates to the underlying PNS changes. We've also tried to send subjects who aren't meditators into the public projects so they have comparison data and can start, at least privately, thinking about the differences.

Ben

But how about people who aren't enlightened masters or regular meditators, but still get into amazingly great states of mind from time to time. Do you think ordinary folks can sometimes get into the same states of mind that characterize the "ongoing peak state" of PNS that you're talking about?

That is, I wonder to what extent these enlightened states are totally beyond everyday peoples' experience, versus just being something that ordinary people can't hold onto after they reach it?

Jeffery

This is a very difficult question. In our research we chose to focus on individuals with persistent forms of the experience. There has been research into temporary forms of it, but we haven't focused at all on that. I don't think we really know how accessible this is within the general population at this point. It may be that certain genetics variations are required for it, for example. We estimate that perhaps as many as 1% of the population in the developed countries we study have a persistent form of the experience. It could be that there is something special about these people.

However, I would say that we've done a huge amount of psychological, lifestyle, etc. research into them. For the most part they are a cross-section of the general population. This seems to occur across all ages, socio-economic levels, education levels, and so forth. Our sample includes people who were devoutly religious and completely atheist. So, there may be some hope for all of us!

I should note that there are a few biases in our sample. It is about 70% male, 98% white, and highly educated. We've tried several sampling procedures to collect new research populations to try to correct these biases. I spent last year in Asia, and it became clear during that time there are plenty of Asian subjects. We started in the US, Canada and Europe so our initial racial bias wasn't too surprising. The gender and education biases have held up no matter what we do to try and eliminate them, though. So there might be something to them in relation to your question. Overall we have around 1200 subjects in our research participant database, and we're not adding to it. We have a backup list of potential subjects that is quite a bit larger.

Ben
You've mentioned a bias favoring research on regular meditators, versus general research on extraordinary well-being and PNS. What do you think are the views on the sort of research you're doing in the mainstream neuroscience community, overall?

Mikey
Academia is just starting to scratch at these areas with the surge in meditation research, but they are still refusing to talk about the reason why meditation was invented. It's not for reducing stress, or depression or anxiety, rather, these are byproducts of a much deeper, more profound shift which is available. It's wonderful to use meditation as a tool in whatever way it can benefit humanity, but a fundamental part of the picture is being ignored. One of the reasons it's avoided is because the nature of the experience is characterized by its sharp contrast to our normal reality. It's just

really hard/impossible to relate to, explain, and understand. This is in contrast to something like depression or anxiety which can be described within the normal range of human experience.

Another problem is that the phenomenon resists scientific verification, largely for historical/cultural reasons. Because the experience is often deeply embedded within a religious or spiritual belief system, it can be hard to talk about the underlying phenomenon. But Jeffery and others have done great work moving past this, and there is no reason PNS couldn't be studied and characterized much like other psychological conditions. But like anything, science can never say 'what' PNS is, it is only one lens through which we can conceptualize it.

Jeffery

I largely concur with this. Ultimately there's only a handful of scientists working in this area, and it is largely outside the system. Funding is generally from interested individuals rather than institutions. There are no professorships, tenure track positions, significant grants, etc. Generally the things that would make a young academic feel like this with a path to a good future career are all absent, and that is a problem. We've worked hard to make some progress in this area but there is still a lot that remains to be done. Where I may differ with Mikey is that I have a slightly more physicalist approach to it all at the present moment. I think it can be quantified in standard science and we're working hard on that.

Ben

On the neural side, what would you say are the most important, exciting results found so far about the neural basis of "extraordinary states of well-being"?

Jeffery

The initial data has converged around default and task mode networks. These are two networks of brain regions that usually activate at opposite times. In other words, when the task network is active generally the default mode is less active. We call this anti-correlation. The task network is generally more active when you are working on a specific task, hence its name. The default network is more loosely defined at the moment, in terms of its functions, but seems to relate in part to an individual's underlying sense of self. It is often active, for example, when someone is just sitting around and not really doing anything. Of course in those moments our mind is often wandering, and this often relates to things that have to do with some aspect of our sense of self. One interesting findings is that in the individuals we research these networks are less anti-correlated.

Ben

That's incredibly fascinating, yeah. Can you say a little also, about the specific brain regions that are most implicated in the different functioning of these networks in "enlightened" people's brains? Of course I know the brain is all about networks and complex circuits, not so much about localized functions specific to particular brain regions. But still, it's interesting to know what regions are most centrally involved in each mental phenomenon....

Jeffery

Within the default network there are two or three areas that seem to be consistently indicated as related to this phenomena. These include parts of the medial pre-frontal cortex (mPFC), the posterior cingulate cortex (PCC), and the insula. The one that has received the most ongoing research to date is the PCC. Much of the work going on in this area is underground, but everyone's findings are very similar so the handful of publicly available work is plenty for people to pursue right now. On the scanning and neurofeedback side, your readers can find out more by looking up work by Zoran Josipovic at NYU, Jud Brewer at Yale, and Kalina Christoff at the University of British Columbia for example. There has also been work by us and others on directly stimulating these brain regions using electromagnetic fields and direct current. And there is quite a bit of parallel work, such as Katherine MacLean and her collaborators work on psilocybin that seems to relate to these overall findings. She is at

Johns Hopkins . Overall it is a very exciting time with a lot of interesting data converging. We're very privileged to be the organization that is talking to everyone and helping to put the pieces of the puzzle together.

Ben

What are your concrete plans for moving your research on neurofeedback forwards? Where is this going? Can we expect to see a "neurofeedback enlightenment X-box plugin" in the Toys R Us next year?

Jeffery

I hope so! Certainly we're working hard on it. We are working on both the neurofeedback and neurostimulation fronts, and literally pursuing every avenue that looks promising. The neurofeedback path is being explored by others as well. For example, Jud Brewer's group initially explored affecting the PCC with real-time feedback during fMRI, and is now looking at how to translate that work into less expensive technology like EEG. We have a similar initiative. To really get it down to the "Toys R Us" level requires a great deal of effort and advancement within the fields we're working in. It has taken us about six years to get to the point where we could begin a serious effort in this direction, so really we are still in the early days. Having said that, we've already had some success with very simple neurostimulation devices being able to induce temporary periods that are highly similar to the persistent states we research. So progress is being made!

Ben

Any thoughts on neurofeedback and "mind hacking"? Is this sort of research something someone can do on their own, via a DIY science approach? From our prior conversations I get the feeling you think the answer is "'Hell, yeah!"

Mikey

Absolutely! The term I've been throwing around is 'Consciousness Hacking', because much of what we're talking about is, at least traditionally, considered outside or beyond the mind. Mystics, monks, and so many others are the original consciousness hackers. They used 'technology' like meditation, breathing, dance, ceremony, etc. to tap into altered states of consciousness, and using those tools they systematically explored the nature of subjective experience. Those techniques didn't appear out of thin air, they were evolved through trial and error for a very specific purpose. And they continue to evolve! To make a categorical distinction between 'traditional' practices such as meditation, and technology assisted practices is ultimately arbitrary, like the lines on a map. They are all human inventions for the same purpose: the cessation of suffering.

That said, I think there is a hugely unexplored and fantastically ripe space for developing tools and techniques to shift our subjective experience and many of these are accessible to the DIY crowd. Neurofeedback and biofeedback have been around for decades and were one of the original forms of modern tech used to shift consciousness. Like much tech, EEG in its early days was inaccessible to the masses, but has now decreased in size, complexity, and price. You can pick up a Neurosky or Emotiv headset, both of which have accessible APIs, and get right to it. Another tech which has experienced a recent surge in popularity is Transcranial Direct Current Stimulation (tDCS). This is a relatively simple device that runs a very small direct current through the brain and can result in distinct, and sometimes significant shifts in experience. The idea of running electrical current through the brain probably deters most folks and that is probably a good thing. Though the technology as its intended to be used is extremely safe, used incorrectly it can obviously be dangerous. There is a wealth of DIY info out there but much of it is rubbish so be very careful.

These were only a couple examples, and there is much more, but unfortunately there is no centralized resource for this info I can point you to. We are working on that, but for the moment anyone interested in this will have to be motivated by the pioneering opportunity it offers. Although there is a small but rich history of technology being used to explore consciousness, very few have looked directly at engineering what Jeffery calls Persistent Non-Symbolic Consciousness. I believe this area of exploration is filled with low hanging fruit, and a community of dedicated consciousness hackers can make real, meaningful discoveries.

Ben

How do you compare the promise of neurofeedback based approaches to achieving extraordinary states of well-being, versus other approaches out there?

Mikey

EEG neurofeedback is sort of the classic technological approach to inducing altered states of consciousness (much to the chagrin of many neurofeedback practitioners), but I think it's only the tip of the iceberg. Recent work in real-time fMRI feedback has been very exciting, showing that people can consciously modulate activity in the areas of the brain that seem particularly relevant such as the PCC and mPFC. And other technologies with better temporal resolution such as MEG, and real time spatial imaging using EEG (e.g. LORETA) are extremely promising. Neurofeedback has the quality that in principle any change enacted through feedback, could be accomplished independently. This is perhaps both an advantage and a limitation. Any technique/technology can be used irresponsibly but it would seem that in the case of neurofeedback the body is directly regulating the activity and thus there is some safety barrier by virtue of natural intelligence.

This is in contrast to brain stimulation technologies which have the capacity to directly modulate brain activity in ways not otherwise independently achievable. Some of these, such as electrical stimulation (tDCS, CES, etc) and TMS, work diffusely and seem to be relatively safe, but lack the precision of something like fMRI feedback. Other technologies such as pulsed ultrasound have great precision and the ability to target almost arbitrary brain regions, and seem to hold great promise. If these states can be induced by modulating activity in a small number of defined brain regions then one of these brain stimulation technologies might be ideal.

But if the experience is mediated by a much more complex and dynamic set of brain networks then a higher-level type of stimulation, such as interactive media, might be appropriate. Games, for example, could be a powerful tool for altering the behavior of the mind. The most effective methods though, may very well lie in the combination of these technologies.

Ben
Any thoughts about even more radical approaches than neurofeedback? What about brain implants that directly measure and modulate the brain appropriately? Wouldn't that ultimately be even more effective than neurofeedback? Would you insert such a plugin into your brain? Want to be the alpha tester?

Mikey
Although I'm all about putting myself out there in the name of science, I'm not sure if I'm ready for experimental neural implants. But, the ability to generate a sufficiently accurate model of an individual's brain might allow us to develop unconventional but extremely effective methods catered to that individual. Imagine a scenario where we could both predict the effect of any stimulus on neural activity, and compute the target brain/physiological state (i.e. what that individual's enlightened brain looks like). Given a set of human inputs and their unique parameters (magnetic, electrical, sound, light, smell, etc.) we could calculate an optimal combination for achieving the desired outcome. For example, that might be a few precise pulses of magnetic stimulation on the skull timed to the rhythm of that individual's brain networks. Or it could be a synthetic voice verbalizing a seemingly obscure series of words in a cadence no human could replicate, transmitted through headphones over a period of weeks. This capability is probably part of any transhumanist vision and would be sufficient to produce a wide variety of subjective experiences, not just transcendent.

Jeffery

Personally I'd love to be a beta tester. I'm looking forward to these types of human enhancement and think they will have a profound impact on all of us. Certainly our area of research could be one of those that benefits from this type of technology, though I hope we have a very simple and effective solution to how to induce this long before brain implants become a common reality!

Ben

To wrap up this long and fascinating interview, I'll turn the dialogue a bit toward some of my own research interests. I'm curious what you think: will AIs ever be capable of enlightenment?

Mikey

An awesome question, and one I definitely do not have an answer to. But I'll throw some thoughts out there.

An obvious barrier to a straightforward answer to this question is the lack of an accepted scientific framework for enlightenment in natural humans. If we can't characterize the phenomenon as it arises presently, it will be difficult to do so in an AI form. But this only pertains to satisfying some ultimately arbitrary conceptual definition. Enlightenment like any other subjective experience, can only be described, or symbolically represented. For example, I can never convey to you directly my experience of sadness, or the color red. We can never say 'what' it is in any absolute way beyond transmitting the experience itself.

Imagining we had some definition, I think there is absolutely no problem with an AI satisfying some kind of spiritual Turing test. Humans frequently manage to deceive themselves and others of their degree of enlightenment. I would even go so far as to say that a sufficiently advanced intelligence would naturally come to the same conclusions found ubiquitously at the heart of eastern philosophy. It's my view that our personal, mentally constructed, dualistic model of reality and self are fraught with paradox and contradiction which any sufficiently deep introspection would reveal.

On a practical level I wouldn't be surprised if early AI systems advanced enough to contemplate their own nature continually thought themselves into unresolvable loops. The designers of these AIs (Ben Goertzel excluded of course!) might not have contemplated the nature of self to its enlightening conclusion and thus not accounted for the dramatically different experience of reality available outside of their own. They might be initially perplexed by the seemingly paradoxical conclusions reached by their creation (SciFi short story pending).

But that of course leaves us with the heart of the question unanswered: "Would that AI really be enlightened?" It is surely possible for people to conceptually realize their nature as fundamentally nondual without coming to that direct experience. It is also possible to for individuals to enter into that direct experience without any conceptual framework. So, there remains this notion of the direct experience of enlightenment, what Jeffery calls Persistent Non-Symbolic Consciousness, which is very specifically not a conceptual knowing or state of mind or mood, but a transcendent state of being beyond description. Can an AI realize that experience, beyond its fundamentally logical/symbolic nature? Depending on our terms this line of questioning quickly propels us into the science and philosophy of consciousness where similar questions are debated ad nauseum. I think this comes to a question of what our fundamental nature is. What is that awareness which is universally and persistently the underlying essence of our experience as humans? It is only by understanding that question, that we can understand if an AI would be able to do the same. So, perhaps like many questions worth pondering, there is no immediate answer, but the process of inquiry leads one closer to truth?

Jeffery

Allan Combs and I held a session on this at the Convergence 08 conference, and it was a fascinating discussion. I suspect that a great deal of this experience is substrate dependent. In other words, I think it is heavily dependent on the human body. While it may be possible to simulate with AI, I personally wonder about what other forms of conscious experience will be possible when the human body is out of the loop. While I believe that the levels of well-being we are researching now are about as far as humans can go with current biology, I doubt that is the case when we start bio-hacking more and I think it is highly unlikely if we change the substrate altogether.

Ben

As you know I'm really interested in the prospect of a technological Singularity. What are your thoughts on this, and in general on the possible role that your own work on neurofeedback and extraordinary states of well-being may play in in the next decades/centuries — as technology advances and further transforms society?

Mikey

I'm very excited about the prospect of Singularity, but I do see some contradictions in terms of the enlightenment experience, and the way in which the Singularity is often talked about. Traditionally PNS is experienced as a state of no self, in which the ego (that which most of us identify as being who we are) is seen as illusory, and ultimately the 'self' is experienced as formless, non-conceptual awareness.

When we imagine our Singularity future in which we transmute into an alternate persistent form, that part of us that we imagine persisting is exactly that which the mystic works to transcend. So, a common Singularity dream of 'living forever' is arguably a very advanced effort to endlessly hold onto our beliefs, desires and the mentally constructed notion of self. From an eastern perspective, those beliefs/desires can be seen as the root of suffering, so in that sense what we are trying to take with us is the root of suffering itself. But, we're simultaneously pursuing a brute force approach to the cessation of that suffering. The strategy seems to be an information/technology infrastructure that supplies the instantaneous satisfaction of all desire. This is in essence the polar opposite of the classic approach to end suffering which is a complete acceptance of reality as it is in this moment perpetually. Perhaps like so many fundamental dualities they are two sides of the same coin?

The enlightenment experience is characterized by a dissolving of the boundaries that separate 'us' from 'other', and the realization of a unitive experience in which all perceived separations are seen as purely conceptual and illusory. Perhaps the Singularity can be seen as the most profound and complete realization of that inherent connection between all things and the fundamentally singular nature of reality. I recently re-read Kurzweil's The Age of Spiritual Machines in which he talks about individuals literally merging their experiences at will into a single perspective. Imagining that all experience is instantly available, and all boundaries between individual's own experiences becomes clearly artificial how do we define our unique identity separate from the whole?

Ben

Great thoughts! I find there aren't many folks one can talk to about things like the intersection between Singularity and enlightenment, because the world seems to mainly fall into

- folks who think techno-Singularity is coming but spiritual stuff is BS
- folks who like spiritual stuff but think techno-Singularity is BS
- folks who think both spiritual stuff and techno-Singularity are BS
- folks who like both techno-Singularity and spirituality, but think about both in verrrry fuzzy ways (e.g. Terrence McKenna, Timothy Leary)

Folks who are willing to think carefully about techno-Singularity and spirituality, without getting carried away in various sorts of wishful thinking OR various sorts of reflexive negative skepticalness, are pretty few and far between... So you guys are really a lot of fun!

Mikey

Perhaps the Singularity demands a fundamental shift not only in the appearance of things, but in our fundamental perception of reality. It might be that our work, and so much like it, is helping to pave the way for that by eventually increasing the accessibility of those transcendent states.

Jeffery

I agree with what much of what Mikey says here. I did a recent interview with Adam Ford that involved this question and the previous one after speaking at the H+ conference in Hong Kong last year. You can find it on YouTube[197]...

Ben

Yeah – that was a great interview. Relatedly, I'll point out to readers a little fictional interview on a related theme recently, that was based on some stuff in an email dialogue between you and me and Mikey not long ago – THE UNIVERSE IS CONTINUING[198].

In the perspective I give there, an "enlightened" view on the Singularity is roughly that it's just another aspect of the universe's mysterious, trans-human play of patterns... Which looks like a major upheaval and development from the limited human view. And we ordinary humans are terribly concerned about the risks and dangers and promises that Singularity holds,

---

[197] http://www.youtube.com/watch?v=-Wt9cBJX8Ww

[198] http://multiverseaccordingtoben.blogspot.hk/2012/07/the-universe-is-continuing.html

only because we're so attached/addicted to our own selves and our lives.

Jeffery

Thanks! I passed that along to my partner, who has essentially the same philosophy.

Mikey

I think we are totally on the same page in this regard. It's hard for me to see a fundamental difference between me typing this email and the leaves blowing on the tree outside. I can conceptualize the difference, but it all seems to be a fantastically complex dance in which all components are completely, utterly, intertwined. To apply any more or less significance to things is fine, but that can only emerge from an initial false notion that those things are somehow distinct and separate from a whole. But in the same way evolution produces increasingly coherent and synchronous forms, the same phenomenon seems to happen with human experience. Suffering seems to be a manifestation of conflict, or disharmony, while increased acceptance of what is (reduced conflict) seems to result in less suffering. That experience of harmony with what is seems to be some kind of experiential direct understanding into the more fundamental nature of reality (It doesn't mean that people with that understand sound smarter. That understanding is ineffable).

So, its pretty fun to try to push on that through tech or whatever — but in the scheme of things, that effort is no different than a house plant growing to be closer to the window.

Ben

Well – this has certainly been one of the more fascinating interviews I've done. Thanks a lot, guys (or should I say — "thanks a lot, fellow hyper-intellectual house plants"?). I look forward very much to seeing how your thinking develops – and hopefully trying out some of the neurofeedback or brain stimulation technology you've talked about!

CPSIA information can be obtained at www.ICGtesting.com
Printed in the USA
LVOW10s1610030614

388429LV00021B/1407/P